BIOLOGY TOPIC TEXT C: CELL REPLICATION AND PROTEIN PRODUCTION

HAROLD M. ICKES II, M.D.

Executive Editor: Rhett C. Smith, Ph.D.

Ideal Quill Publishing Group, LLC

idealquill.com

Printed in the United States of America

10 9 8 7 6 5 4 3 2 1

ISBN 978-0615834849

Biology Topic Text C: Cell Replication and Protein Production

by Harold M. Ickes II, M.D.

Table of Contents

Unit 1

Replication

In this section we cover:

1) DNA Replication

2) Differences between Eukaryotic and Prokaryotic replication

In the earlier topic texts, we talked about the stuff from which life is made, and then we got our feet wet with a little biochemistry when we discussed how life gets its energy. Now, we turn our attention to life's ability to reproduce. DNA is at the center stage of this particular drama, so let's focus our attention back on this information molecule to get us started on the discussion.

As you'll recall, DNA is a long chain of nucleotide monomers. There are four different kinds of nucleotides: A, G, T and C, shown here:

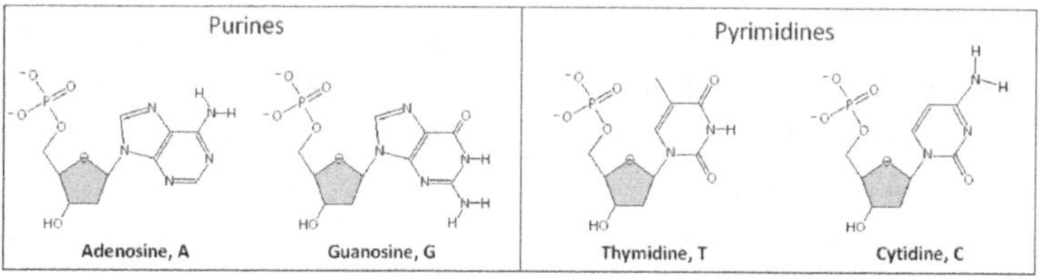

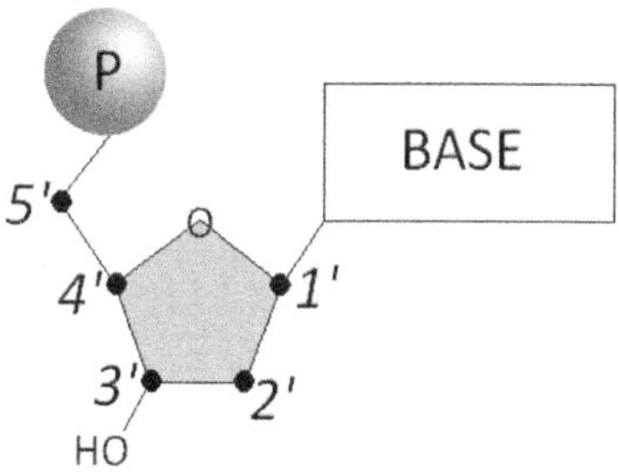

The carbons in the gray sugar ring are numbered from 1' to 5' (read as "one prime" to "five prime"). The phosphate of the nucleotide is attached to the 5' position.

When nucleotides are strung together to make DNA,

the numbering system helps describe the directionality of the polymer that results.

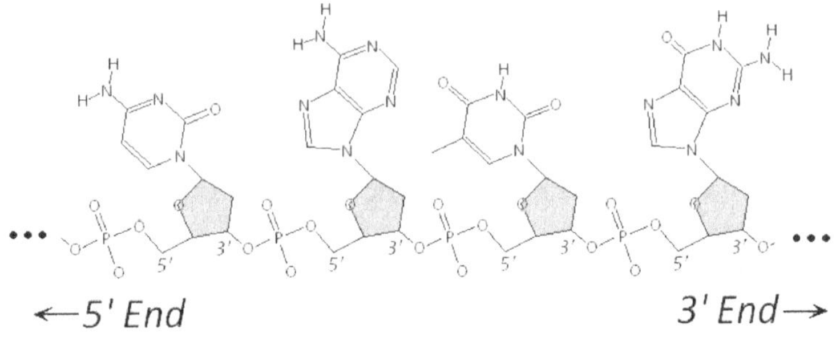

Each chain of nucleotides has a complementary chain that attaches to it, side-by-side, like a long ladder, each rung being a nucleotide-pair. The opposing strands of DNA run in opposite directions. One strand goes from 5’ to 3’ while the other is 3’ to 5’.

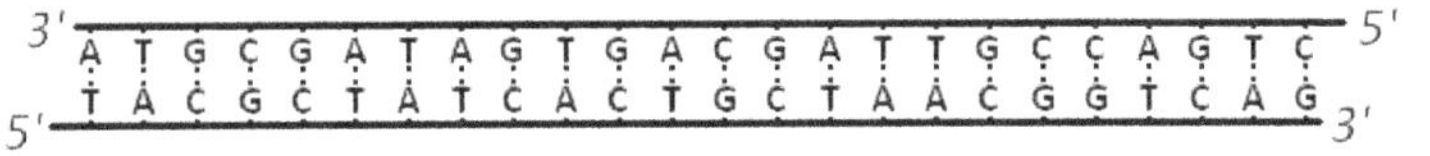

In addition to opposing directions of the strands, the nucleotide pairs have a very special nature. If an A is on one side, then a T is on the other. If a G is on one side, then a C will be on the other. This special interaction occurs through hydrogen bonds. A and T share two hydrogen bonds, while G and C share three. That said, this DNA “ladder” is also twisted, resulting in a complex three-dimensional shape called a double helix.

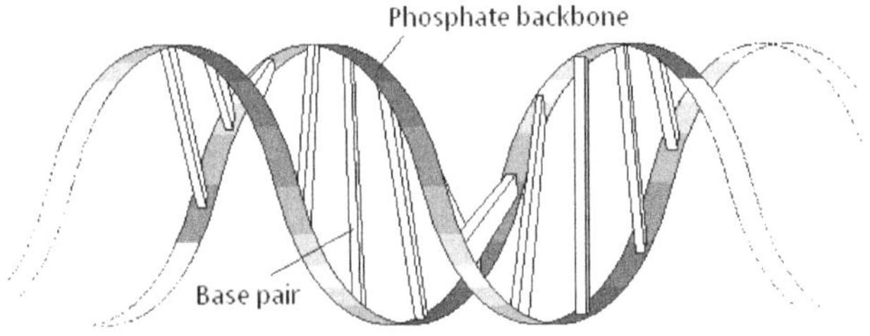

The DNA helices of living organisms typically contain hundreds of thousands of these nucleotide pairs. In bacteria, these long DNA ladders are circular with the ends attached to one-another, while in eukaryotes, such as humans, the DNA has two free ends (called **telomeres**). The difference between prokaryotic and eukaryotic DNA goes a little further, though, because eukaryotic DNA is also wrapped tightly around protein clumps called **histones**. We touched upon this briefly in our chapter of cell anatomy in Topic Text B. Five histone proteins are connected together to make a larger protein complex, and the DNA is wound around this complex. Taken together, the entire thing is called a nucleosome (see unit 6, Topic Text B). Either way, we call one full chain of DNA, (whether looped or straight), along with whatever proteins that help package it, a **chromosome**. Bacteria typically have only one chromosome. Eukaryotes, on the other hand, differ in the number of chromosomes they have, although their chromosomes generally come in pairs. Humans, for example, have 46 chromosomes, or 23 pairs. The complete set of chromosomes that an organism has is called its **genome**. So, most bacteria (for simplicity) have one chromosome in their genome, while humans have 46 chromosomes in their genome.

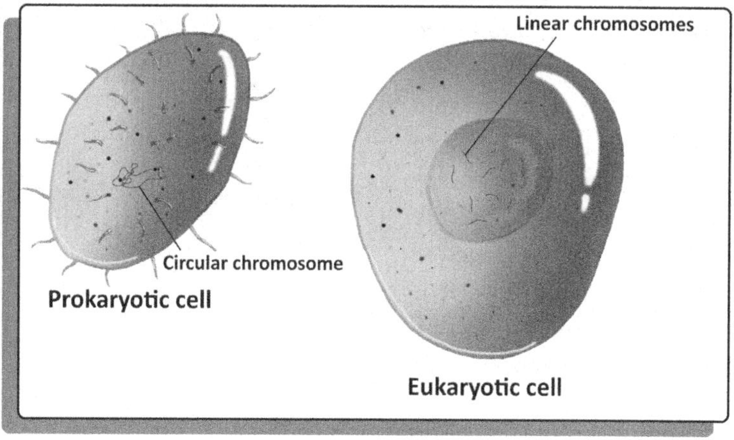

That said, let's talk about the complementary nature of the double helix strand some more. Suppose we are out walking around and happen to stumble across only one strand of a DNA helix. The complementary strand is missing for some reason, so that the strand we find looks like half of a ladder. Because of the interesting complementary nature of DNA base pairs, we already know what the other half of the ladder should look like! Wherever there is a G, the other side will have a C. Wherever there is an A, the other side will have a T, and so forth. In other words, one side of the DNA helix contains the information for the other, and vice versa. This fact is important in the replication of DNA. As a cell prepares to divide, it slowly pulls apart the DNA helix and makes a new copy of both halves using the complementary information. When this is done, the cell ends up with two copies of the original chromosome. Life apparently had copy machines long before we did!

DNA Replication

There are subtle differences between how prokaryotes and eukaryotes replicate their DNA, but the complex geometry of the helix imposes the same problem either way. Let's say we separate the helix altogether so that the two halves are floating free in the cell. The two halves are complementary and will want to reconnect (**anneal**), but do you think they anneal just like they were before? Will the two halves come back together perfectly? Theoretically, it is possible, but the odds are actually quite remote. The problem is that any given A doesn't know *which* T it was

originally bound to, and similarly with the G's, C's, and T's. Instead of annealing perfectly, two other events are much more likely to occur: 1) The two halves of the DNA would clump together in a huge tangle, or 2) a strand could double over so that its own A's bind to its own T's, its C's bind to its G's, and so on. Either way, you get a complete mess. The DNA strands in living systems are just too long to be pulled apart entirely.

This leads to a very important rule: No DNA strand can be left without a complementary strand. If the cell needs to work with the DNA base pairs, it separates only a small part of the helix, and then puts it back together after it's done.

This rule remains true for the replication process as well. Replication begins at an origin of replication, where proteins bind and slowly pull a small section of the strands apart. If the cell has to start somewhere on the DNA chain to begin replication, how does it know where in the chain to begin? Well, DNA has specific start sites that typically have a LOT of A's and T's. A-T pairs bond with only two hydrogen bonds, whereas G-C pairs bond with 3. This means that A-T bonding is weaker than G-C bonding. It also means that a long stretch of A-T's will be easier to pull apart than a long stretch of G-C's. So, a collection of proteins called the **pre-replication complex** binds to the origin of replication and pulls this long stretch of A-T's apart so that replication can begin.

After the pre-replication complex completes this work, the copying machinery comes in and begins inserting the new nucleotides to both of the original halves of helix, moving in opposite directions as it does so. In this manner, neither of the original strands is left with long stretches of unpaired nucleotide bases. To see how this all works, let's start with a very simplistic model of the replication machinery, and slowly add complexity to the images as the details emerge.

Our highly simplistic first view of DNA replication.

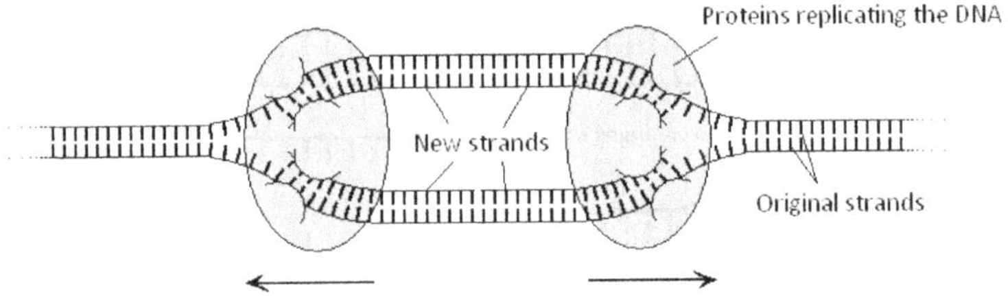

The actual separation of DNA strands is accomplished by a protein called **DNA helicase**, but here we are presented with a problem. Separating DNA strands is not as easy as it sounds. The two strands are wound around each other in a helix. That means if you try to pull the two strands apart, they will still be spiraled around each other and get tangled. In order to fully separate them, you must *untwist* them as well. The following image shows this process.

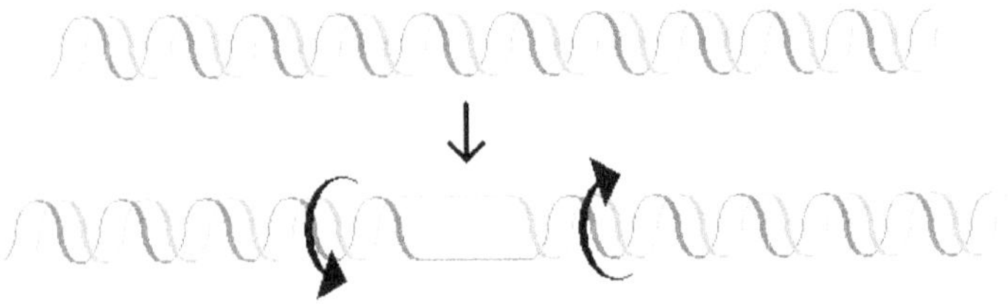

The job of untwisting the DNA helix falls to a protein called **DNA gyrase.**[1] Together with helicase, the two proteins are used to separate DNA strands anytime the cell needs to do work with the base pairs, such as during replication or in the initial stages of protein production (discussed in later chapters).

So once helicase breaks open the DNA at the starting spot and the gyrase untwists the DNA, we are ready to create the new DNA strands from the unraveled single strands. The protein responsible for creating the new strands is called **DNA polymerase**[2]. One DNA polymerase moves along the helix in one direction, while another DNA polymerase protein moves in the other direction. In front of each DNA polymerase are helicase and gyrase.

A better but still simplistic view of Replication

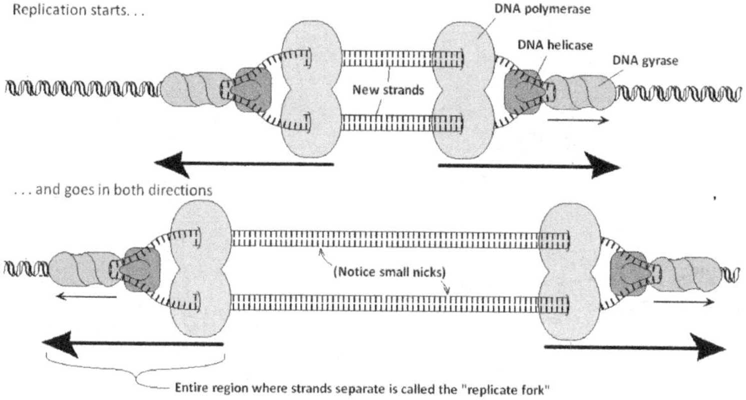

---

[1] Actually, gyrase doesn't technically "untwist" DNA. It cuts the backbone of one strand, spins that strand around the other, and then seals up the backbone again. The effect, however, is the same as if the protein untwisted the DNA. The reason it doesn't simply untwist the helix is that it would have to untwist it all the way to the end, which could be millions of nucleotides away. That would be very slow and tedious work.

[2] Each DNA polymerase is actually made of two circular polymerase "holoenzyme" proteins, each shaped like a donut. The DNA is threaded through the centers, like two strings running through a figure eight.

The last image shows little nicks in some of the backbones of the new strands. To see why these happen, we'll need to dig a little deeper into the process. As it turns out, the generation of the new strands is a little more complex than our initial overview would suggest.

DNA polymerases work by reading the old strand from the 3' to 5' direction. As we saw above, complimentary strands face in opposite directions. So, if the polymerase reads the *old* strand in the 3' to 5' direction, it adds the *new* strand in the 5' to 3' direction. In other words, it adds new nucleotides onto the 3' end of the new strand.

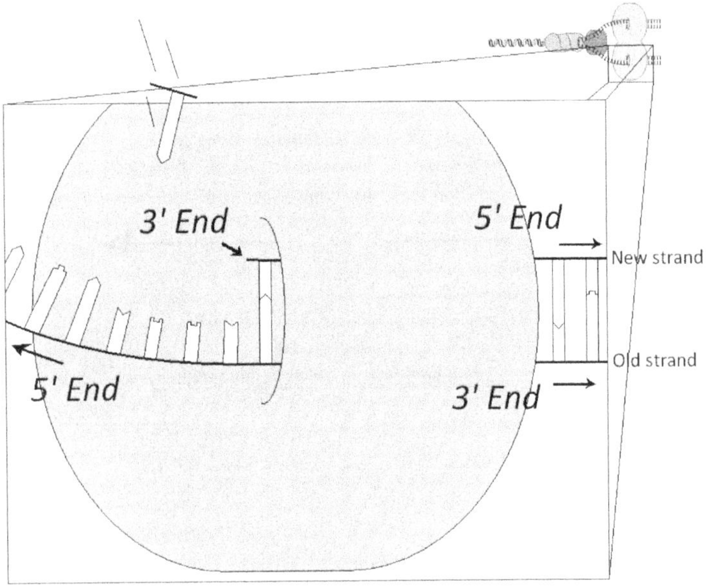

A problem arises because only one of the old strands is facing the 3' to 5' direction. That means if a polymerase is scooting along the helix, one strand will be facing backwards. Since polymerase works by reading old strands in the 3' to 5' direction, it must loop this backwards strand around and read it in the other direction:

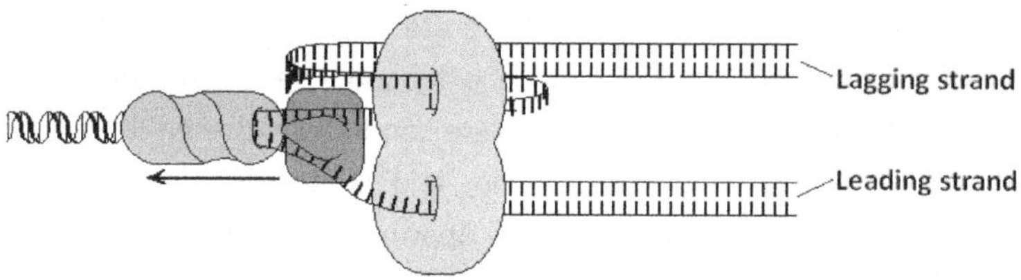

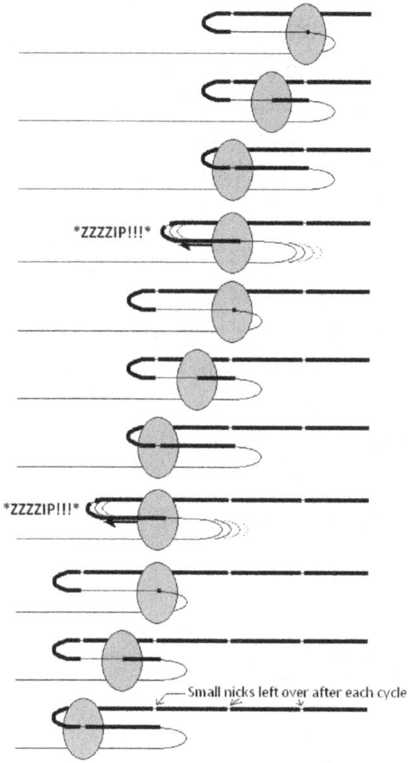

By looping it around, the polymerase is affectively able to read the strand in the same direction as the other strand: 3' to 5'. This strange looping process seems quite bizarre and is rather difficult to grasp the first time it is encountered. Conceptually, it is hard to visualize just how the mechanism would work as the polymerase moves down the helix. To the left is a cartoon to help show the process. In the cartoon, think of the polymerase as a bead and the DNA as a string. The bead slides down the string until it hits the bend in the string, at which point you need to pull more string through the hole in the bead to have enough slack to keep sliding it.

In the image, the DNA polymerase slowly moves left. Just as the DNA polymerase bumps into the looped strand, the strand is threaded forward. The DNA polymerase continues moving left until it nearly bumps into the strand again, and so forth. This gradual ratcheting movement is a reasonable approximation of how it occurs in the cell.

Sadly, there is one more problem that needs to be addressed when it comes to the synthesis of a new DNA from the stubborn 5'-3' looped strand. If you look at the close-up picture of the DNA polymerase several pictures above, you'll note that DNA polymerase adds new nucleotides to the 3' end of the *previous* placed nucleotides. If there are no previously placed nucleotides, DNA polymerase will stall. *This is a problem.* If you look back at the above cartoon, each time the thread ZZZZIPs forward, it appears as though the DNA polymerase has to start a new thick line from scratch. There is one more piece of information, however, before we have the *completed* picture: before the polymerase starts its work on this looped strand, a protein called **DNA primase** lays down a temporary stretch of RNA nucleotides, called an **RNA primer**. That primer gives the DNA polymerase a 3' end upon which to add new nucleotides. With this correction, the individual parts of the above cartoon would look something like this:

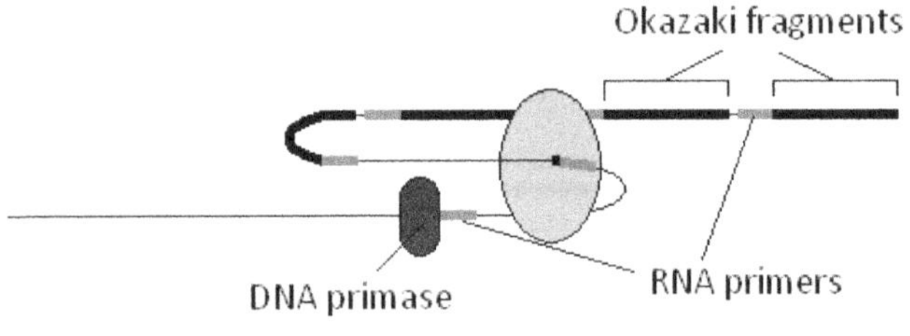

From this process, the new strand that results is a series of short DNA stretches called **Okazaki fragments,**[3] which are separated from each other by small nicks (places where the phosphate backbones are not sealed) and short sequences of temporary RNA nucleotides. To complete the synthesis of this rag-tag strand, **DNA ligase** seals up the nicks in the backbone, and then another type of DNA polymerase[4] replaces the RNA primers with DNA nucleotides.

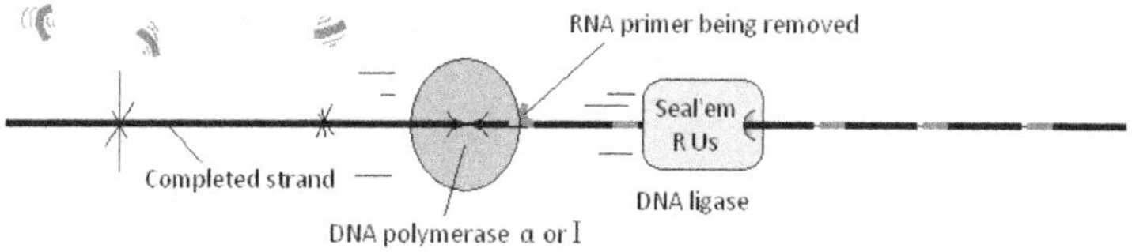

Overview

Let's put it all together. . .

Two DNA polymerase molecules attach to the DNA near the origin of replication, where a set of proteins, called the pre-replication complex, have gathered to mark the spot for replication to start. Once attached, the two DNA polymerases move along the helix in opposite directions. The helix itself is pulled apart and untwisted by helicase and gyrase. As each polymerase moves along, it begins making new strands using the old strands as templates, matching the nucleotides with their complementary bases. The 3'-5' old strand is used in a straight forward fashion, and new nucleotides are added sequentially

---

[3] Although they may appear small in the cartoons of this chapter, Okazaki fragments are actually several hundred to a thousand base pairs long, depending on the species. Eukaryotic Okazaki fragments are usually much shorter than prokaryotic Okazaki fragments.

[4] The protein is named DNA polymerase α in eukaryotes and DNA polymerase I in prokaryotes. Generally speaking, while the patterns of many biological processes are often the same between eukaryotes and prokaryotes, the names and features of the proteins that execute the individual cellular mechanisms can vary quite a bit. Also, in some cases, proteins can have multiple names. For example, DNA gyrase (the protein that untwists the helix) is also known as topoisomerase. It can make learning the material difficult, so keep the possibility of name variation in mind.

without difficulty. This creates the leading strand. The 5'-3' strand is facing in the opposite direction, however, so it has to be looped around the polymerase and threaded forward periodically. Since the polymerase requires an existing 3' end of a previously placed nucleotide, a protein called DNA primase inserts a temporary sequence of RNA nucleotides, called RNA primers. New nucleotides are then attached to the 3' end of these primers. The new strand is thus made in short lengths called Okazaki fragments. Each fragment is separated by little nicks, where the backbones have not yet been bonded together. Finally, after this is finished, DNA ligase seals up the nicks in the backbone, and another type of DNA polymerase replaces the RNA primers with DNA nucleotides. This completes the lagging strand synthesis.

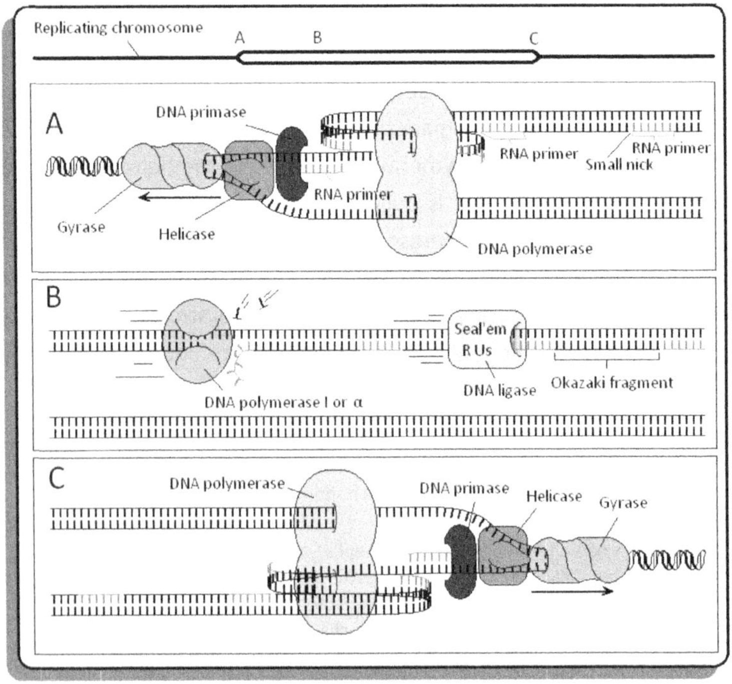

Replication: Prokaryotes vs Eukaryotes

The primary difference in DNA replication between eukaryotes and prokaryotes is in the number of origins of replication used. Because prokaryotic DNA is circular, bacteria typically have only one origin of replication. Replication starts there and races in both directions around the circular track until they meet each other. At that point, each of the old strands has a completely new complementary strand.

Replication of prokaryotic chromosome

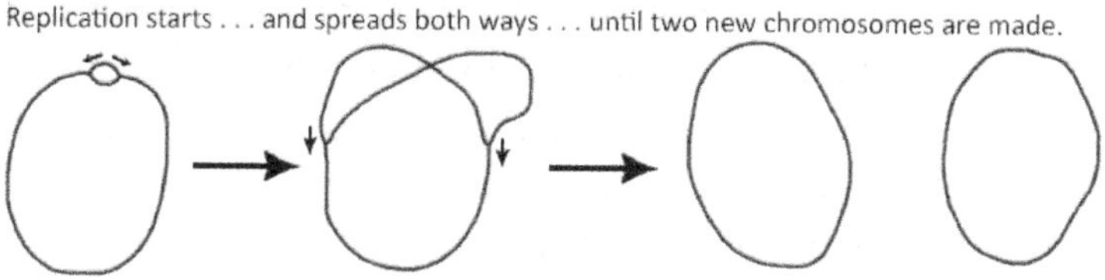

Eukaryotic chromosomes generally have many more than one origin of replication. Why? First, eukaryotic DNA is wound around histones. Supercoils are also present, so that unpacking the DNA is a considerably slower process. Our bodies are choked-full of cells that divide at a rapid rate, so they need to be able to replicate their DNA quickly! That means the DNA needs to be copied from multiple places at the same time, or the process would take FOREVER to get done! Therefore, eukaryotes have multiple origins of replication. One last thing to point out is that the origins or replication are always somewhere within the chromosome and *never* on the ends.

Replication in eukaryotes

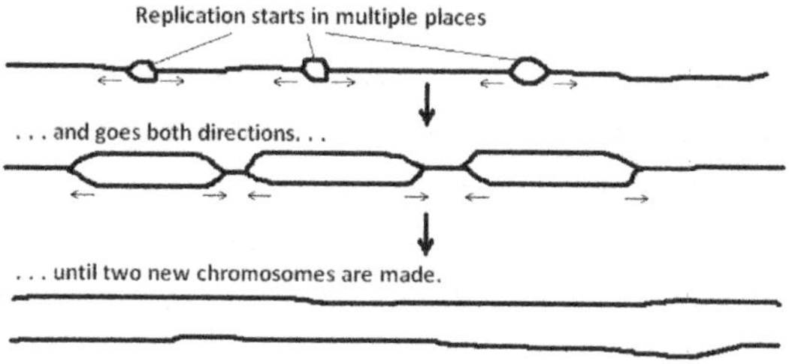

Summary

In this section, we talked about DNA replication.

- DNA is comprised of four different base pairs: adenosine (A), guanosine (G), thymidine (T), and cytidine (C). A's pair with T's through two hydrogen bonds, and C's pair with G's through three hydrogen bonds.
- The strands of DNA are paired up in a helix, which looks like a twisted ladder.
- Anytime the cell needs to access the base pairs, it separates the two strands very minimally to reduce the chance of errors when the strands come back together. The cell never leaves a strand by itself because it will not fit together again correctly with the other strand.
- DNA replication begins at an origin of replication, which is a long stretch of A-T pairs. A set of proteins called the pre-replication complex binds to the origin of replication, marking it as a place for replication to begin.
- Next, two DNA polymerase proteins attach and begin making new strands, using the old strands as templates. A's are matched with T's, and G's are match with C's.
- Two polymerase proteins move in opposite directions to make new strands, like a rope cut down the center.
- The DNA strands are difficult to separate since they form a twisted helix. In order to open the helix, it must be untwisted as well as pulled apart. The job of untwisting the DNA falls to DNA gyrase. The job of pulling apart the exposed base pairs comes to DNA helicase. Together, they move along the helix in front of DNA polymerase, pulling it open so the polymerase can access the base pairs of the old strands.
- DNA polymerase reads old strands in the 3′ to 5′ direction, and makes the new strands in the 5′ to 3′ direction. Only

one of the old strands is facing in the 3' to 5' direction. This strand is paired with a new strand (the leading strand) quite easily. The other old strand is facing backwards. It must be looped around the polymerase and read in the opposite direction. As the polymerase moves along, this backwards strand is periodically ratcheted forward. The result is a new strand being made in fragments, separated by little nicks where the backbone of the new strand has not yet been sealed. This fragmented strand is called the lagging strand, and each new fragment is called an Okazaki fragment.

- DNA polymerase requires the 3' end of a previously placed nucleotide in order to lay down a new nucleotide. Since the looped strand is facing backwards, there are no previously placed nucleotides for polymerase to work with, so a protein called DNA primase puts a temporary strip of RNA down to provide the 3' end. This temporary strip is called an RNA primer.
- As DNA polymerase moves away, DNA ligase runs along the lagging strand, sealing up the backbone. Then, another form of polymerase (α in eukaryotes or I in prokaryotes) follows after the ligase, replacing the RNA primers with DNA nucleotides.
- The process of replication is quite similar in both eukaryotes and prokaryotes, except for the number of origins of replication. Prokaryotes typically have one circular chromosome, with one origin of replication, while eukaryotes have many linear chromosomes, each with many origins of replications. Eukaryotes have more origins of replication because their DNA replication takes longer to replicate; it wouldn't get done in time with only one origin of replication.

Unit 2

Protein Production, Part 1: RNA Transcription

In this section we cover:

1) The Central Dogma of Biology
2) The concept of the Gene
3) RNA transcription
4) Post-transcriptional modification of mRNA

The Central Dogma of Biology

In the last section, we saw how DNA is replicated. Now, let's turn our focus to the primary importance of DNA: protein production. DNA stores protein blueprints. The proteins that are made from these blueprints manipulate the environment to increase the odds that the genome will make it to the next round of replication. In other words, proteins are the genome's tools for survival.

At this point, a nagging question might be playing on your thoughts: how did DNA make protein in the first place if it needed proteins to do it? How did it even replicate to begin with? Isn't this a chicken and egg scenario? Not quite, but that question is a natural one for many students to have. A full appreciation of the elegant answer requires a lot more understanding of biology. By the end of the next topic text, we will be in a position to tackle the explanation.

Let's get back to the story. The important thing about DNA is the *order* in which the nucleotides are attached to each other. This order is used as a means of storing a code. It's a lot like the letters in our language. Using only 26 letters and a few punctuation marks, we can express a limitless number of ideas, (and hopefully one or two good ones). All we have to do is arrange the letters in some meaningful fashion. DNA works in a similar way. Each triplet of nucleotides translates into one of the twenty amino acids, and the order of the triplets appearing in the DNA chain tells the cell the order in which to put together the amino acids. This order of amino acids then determines the structure and behavior of the protein (primary, secondary, tertiary and quaternary structure as discussed in our unit on proteins in Topic Text A).

Not all of the genome is used to make protein though. In fact, it is generally believed that most of the DNA, like our libraries and governments, is full of junk. You might, for example, find a short stretch of DNA that encodes for a protein and then go for thousands and thousands of base pairs that don't encode for anything whatsoever. All at once, you will again encounter another stretch of meaningful DNA, and so forth. It's like a book full of random letters and meaningless drivel, with an occasional sentence here or there.[5] The meaningful little stretches of useful DNA are called **genes**. The rest, quite simply, is called junk.

Like our written language, biological systems use a kind of punctuation as well. Genes are sandwiched between small sequences of nucleotides that tell the cell's biological machinery when to start reading and when to stop. The "start" is a recognizable set of base pairs called the **promoter**, which various proteins bind to in order to start the process of reading the gene. The "stop" punctuation, on the other hand, varies from eukaryotes to prokaryotes. We'll discuss this more later in this unit.

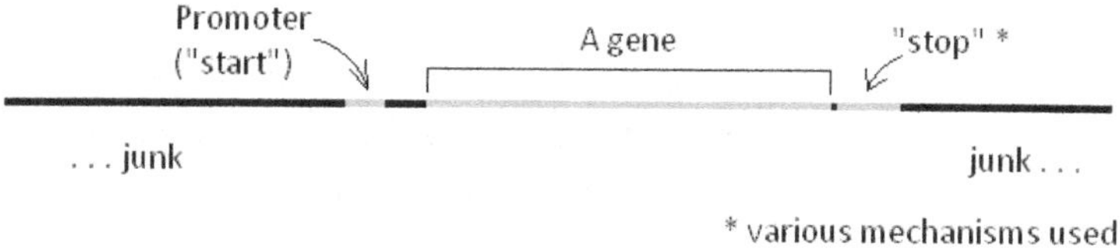

That's all fine and good, but how does the gene get put into action? Well, the first thing that happens is **RNA transcription**, which

[5] . . .just like this particular book, for example.

is a fancy name that basically means a small "copy" of the gene is made using RNA. During this process, the DNA helix is pulled open a little and the strand *opposite* the one having the meaningful base pairs is used as a template to put together an RNA strand, called a **messenger RNA**, or **mRNA**. Since the base pairs bond in a complimentary fashion, the mRNA that results will be an exact RNA copy of the strand having the gene.[6] The image below shows this idea. From this image, you can see that the strand that contains the encoded protein is called the **sense strand**. The strand that is complimentary to the sense strand is the **nonsense strand**.

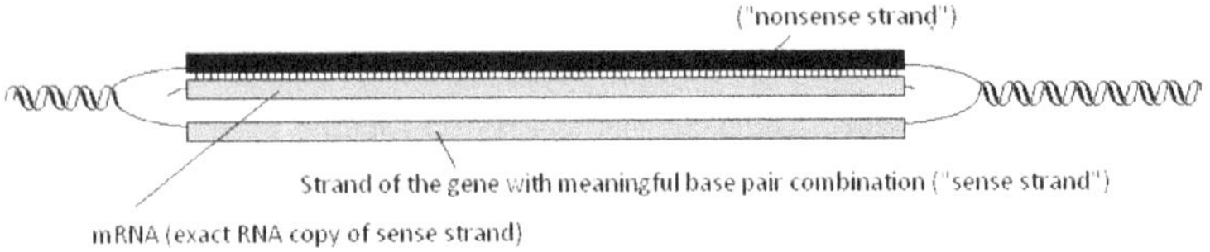

After this happens, ribosomes take these little RNA strands and use them to decipher the order of amino acids needed for the protein in question. Taken together with RNA transcription, this entire process, (DNA to RNA to protein), is dubbed the **Central Dogma of Biology**. For the rest of this chapter, we will cover the first part of the process, the transcription of RNA from DNA. Then, in the next chapter, we'll finish up the story by covering how protein is created from RNA.

RNA Transcription

If you'll recall, RNA is just like DNA. It is made from nucleotide monomers similar to those that make DNA, where the major differences are 1) the sugar unit on the nucleotides in RNA has an extra hydroxyl group (-OH) attached to it, and 2) thymidine is

6 "Exact" is used somewhat lightly here. RNA is technically different than DNA, but the base pair sequence is the same as the sense strand, except that all T's are replaced with U's, as you'll see in the rest of chapter. I use the word "exact" to help facilitate the concept that the RNA transcript is the same functional equivalent of the sense chain.

replaced by uridine. Uridine is very similar to thymidine, but with the methyl group missing, as shown here:

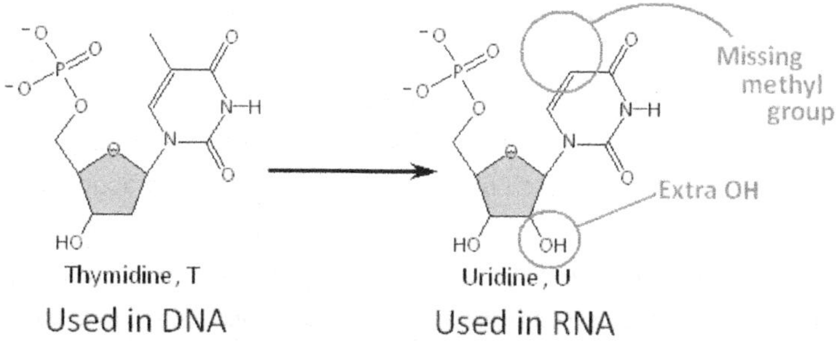

These RNA nucleotides are strung together into strands (i.e., polymers), where the function of the strand depends, in large part, on the sequence of the nucleotides. RNA polymers are divided into many categories. The three largest RNA categories are messenger RNA (mRNA), which we encountered in the previous section, **transfer RNA (tRNA)**, and **ribosomal RNA (rRNA)**. We will encounter transfer RNA in the next unit and will therefore put off discussion of it until that point. Ribosomal RNA are polymers that help make up the two gloves of ribosomes (see unit 6 of Topic Text B). Here is a picture of how the three categories typically look.

Three kinds of RNA

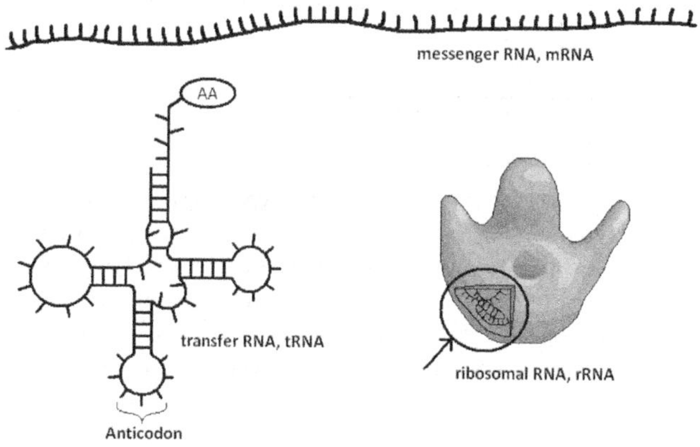

Messenger RNA are the blueprints for proteins and are made in RNA transcription. At the very start of this process, an enzyme complex called **RNA polymerase** binds to a promoter, or "start" signal, sitting just before a gene[7]. Think of the promoter as the landing platform for the RNA polymerase complex. It has a very recognizable pattern, usually something like: TTGACA . . . . . .TATA, where the ". . . ." represents a short stretch of base pairs that could be practically any combination. Once the binding of RNA polymerase to the promoter is complete, the structure begins moving along the DNA. As it does so, it reads the nonsense strand in the 3' to 5' direction, matching up the strand with RNA nucleotides that float by and enter an intake space on the top of the complex. If the RNA polymerase reads a G on the nonsense strand, it attaches a C to the growing messenger RNA. If it reads a T, it attaches an A. If it reads an A, it attaches a U (uridine instead of T, thymidine), and so forth. The growing messenger RNA emerges from a space in the top of the polymerase complex.

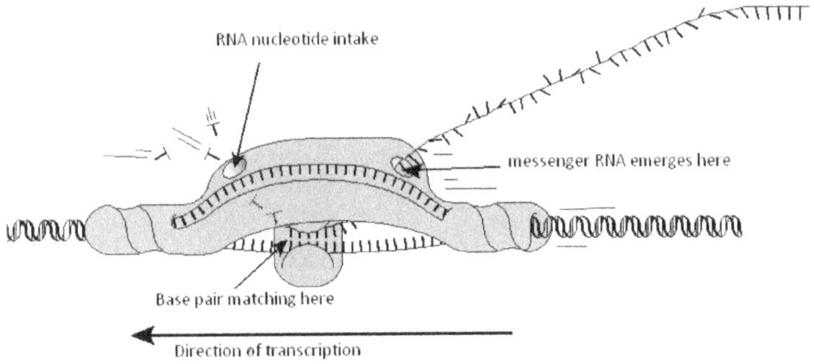

---

[7] There is a great deal of variability in this process, depending on the species, though the patterns are the same. Generally speaking, in eukaryotes RNA polymerase binding happens piecemeal. The proteins that are involved in the complex land separately, gradually forming the RNA polymerase complex. In prokaryotes, the RNA polymerase complex (made of many subunits) is more-or-less already formed when it binds to the gene promoter. Here, we avoid the minutia in attempt not to lose the forest for the trees.

Just as in DNA replication, the problem of unwinding the DNA must be handled, or the DNA cannot be affectively separated without tangling. To handle this, the RNA polymerase complex incorporates DNA gyrase-like and helicase-like subunits that accomplish this job. As it moves along, these subunits separate and unwind the DNA, threading the sense strand to the side and out of the way. As the polymerase passes it winds the DNA back together again behind it. As always, the DNA strands are never left unpaired for very long. All told, it looks like a little bubble moving along a string.

At this point, you may be wondering why the mRNA that emerges from the RNA polymerase doesn't clump together into a jumbled ball. After all, that is what DNA and RNA strands do if left without a complimentary partner strand, and we *did* make a fuss about it in the last unit. The reality is that the mRNA *does* clump up here and there, but it also tends to have a lot of proteins (not shown in the picture for simplicity) interacting with it. These proteins, no matter their other roles, also prevent aggregation (clumping). Also, a small amount of RNA chain aggregation is not nearly as problematic as aggregation of an entire genome would be, so mRNA clumping is a minor issue.

At any rate, this process of mRNA creation continues until the "stop" signal is encountered. In prokaryotes, a stop could be signaled by one of two methods, depending on the gene and the species. One "stop" method, for example, is when a stretch of self-recognizing G-C rich base pairs is created in the growing mRNA. This self-recognizing area creates a hair-pin (an abrupt turn of the chain onto itself) in the mRNA, and this occurs while a stretch of A's are currently being read. As we saw in DNA, A's bind less tightly to their complementary bases than do G's and C's. As a result, the stretch of G-C in the mRNA will snap together and yank the forming end of the messenger RNA out of the RNA polymerase. This event is believed to cause the RNA polymerase complex to pop off the DNA strand.

Example of transcription termination

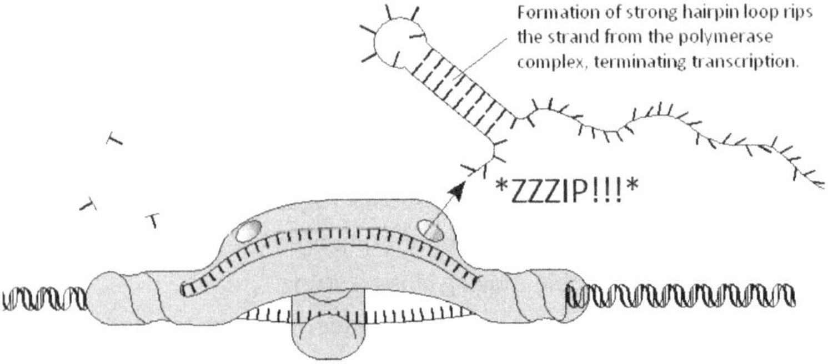

Another way transcription ends in prokaryotes is when a protein called "Rho" attaches to the free end of the new mRNA and crawls toward the RNA polymerase complex. When it reaches the complex (where the mRNA is emerging from the polymerase), it causes the process to stop. How RNA transcription ends in eukaryotes, such as in you and me, is not fully understood.

After transcription termination, the fate of the mRNA differs in prokaryotes and eukaryotes. In prokaryotes, the mRNA proceeds more-or-less uninterrupted to the protein-making part of the process. In eukaryotes, the messenger RNA, or **pre-mRNA** as it is called at this stage, experiences a number of modifications before it is ultimately ready to serve as the protein blueprint. The first two changes that occur are 5' "capping" and **polyadenylation** of the 3' tail. The 5' end of the messenger RNA is capped with the addition of **7-methylguanosine**, which is basically an altered guanosine snapped backwards onto the front of the mRNA. Meanwhile, a stretch of 100-200 adenines is attached to the 3' end. This is called the **poly-A tail**.

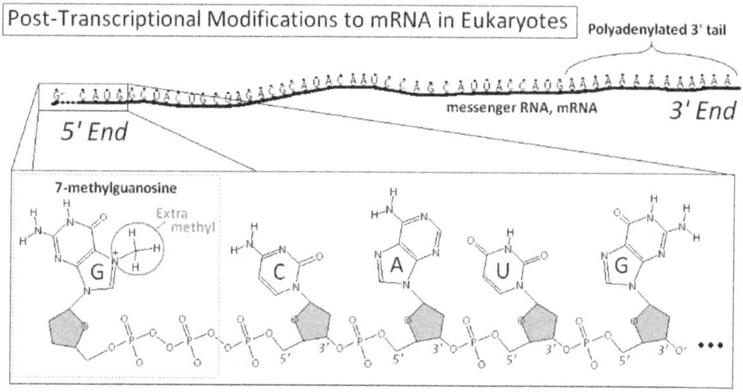

After these modifications, short stretches of base pairs within the pre-mRNAs are often removed. It's like taking a string, snipping out small parts in the middle and gluing the rest back together. The pieces that are removed are called **introns**: they stay IN the nucleus. The pieces that are not removed are called **exons**: they ultimately get to EXIT the nucleus. This process is called **intron splicing**, and it happens when small RNA-protein complexes called **snRNP** (read as "snurps") recognize specific sequences in the mRNA. These snurps also combine with the other proteins and, collectively, the entire mess is called a **spliceosome**. After the formation of the spliceosome, the introns are removed (as **lariat loops**) and the ends of the exons are spliced together. After all the introns are removed, the pre-mRNA becomes a mature mRNA and is finally allowed out of the nucleus to play.

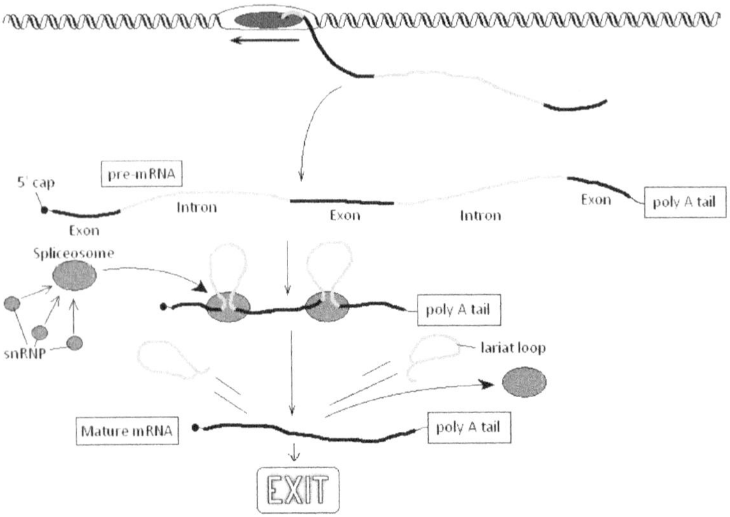

With that, we bring to a close the first half of the story of protein production. In the next unit, we'll finish the story and see how the mRNA interacts with ribosomes to make protein.

Summary

In this section, we discussed RNA transcription.

- The general theme of information flow from DNA to RNA to protein is called the Central Dogma of Biology. It is divided into two parts, occurring in order, RNA transcription and RNA translation.
- Most of the genome in any given species is full of random nucleotide sequences, called junk. Small stretches here and there contain the information for proteins. These small stretches are called genes.
- Genomes that contain the information for efficient proteins will be more successful in surviving to replication than genomes encoding for crummy proteins. Over time, the more efficient genomes will outnumber the less efficient ones.
- Genes have "start" signals in front of them called promoters. They also have "stop" signals as well, but these appear in a number of different variations.
- The process of protein production starts with RNA transcription.
- RNA transcription begins when a complex of proteins called RNA polymerase binds to a promoter. Once fully formed, the RNA polymerase moves along the gene, making mRNA from the nonsense strand.
- In prokaryotes, transcription stops when either a G-C hairpin loop forms and yanks the rest of the mRNA away from RNA polymerase or when a Rho protein crawls along the mRNA and reaches the RNA polymerase. Transcription termination in eukaryotes is not well understood.
- In eukaryotes, the mRNA are first known as pre-mRNA and experience a number of modifications before becoming a

mature protein blueprint. In prokaryotes the mRNA proceed to the next step without any modifications.

- The first modification to pre-mRNA is 5' capping with a backwards-facing, altered guanosine called 7'-methylguanosine. The second modification is the addition of 100-200 adenines to the 3' tail of the pre-mRNA. This is called the poly-A tail.
- Lastly, a number of introns are removed from various points along the mRNA. These are removed from the mRNA by spliceosomes. The remaining mRNA pieces, called exons, are bonded together, resulting in the mature mRNA.

Unit 3

Protein Production Part 2: RNA translation

In this section we cover:

1) The Genetic Code
2) transfer RNA
3) RNA translation
4) Protein routing to endoplasmic reticulum
5) Speed of protein production

In the last chapter, we discussed the first half of the Central Dogma of Biology, where messenger RNA is transcribed from DNA. The second half of the dogma is called **RNA translation** and is the focus of this chapter. It has this name because the cell's machinery takes the mRNA and translates the order of its nucleotides into the order of amino acids for the new protein. Before we tackle that process, however, let's talk for a bit about the genetic code as well as about transfer RNA (tRNA). An understanding of both is vital before moving on.

The Genetic Code

We spoke a bit in previous sections about the information that DNA holds. Let's zoom in a little more into how this works. It's all based on *combinations*. Consider a stretch of 3 nucleotides (a triplet). The first nucleotide can have one of 4 bases, so we start with 4 possibilities: U, A, C and G. Similarly, the second nucleotide can also have 4 possible bases, so together there are sixteen (i.e. 4 × 4) combinations available: UU, UA, UC, UG, AU, AA, AC, AG, CU, CA, CC, CG, GU, GA, GC and GG. After the third nucleotide, our total combinations increase to 64. In other words, since each of the three nucleotides can be one of four possible bases, there are a total of 64 possible triplets we can make.

So what, right? Well, the cell has only twenty kinds of amino acids. The trick living systems use goes as follows: they assign one amino acid for each triplet (hereafter called a **codon**), like say, AAA is the code for lysine, GGA for glycine, UAC for tyrosine, and so forth. With 64 possible codons, there are more than enough combinations to cover the twenty amino acids. In fact, there are so many codons left over that multiple codons can be assigned to a single amino acid (i.e., GGU, GGC, GGA and GGG all go to glycine). Here is the full list of the codon assignments (called the **genetic code**)[8].

	U		C		A		G	
U	UUU	Phenylalanine, phe	UCU	Serine, ser	UAU	Tyrosine, try	UGU	Cysteine, cys
	UUC	Phenylalanine, phe	UCC	Serine, ser	UAC	Tyrosine, try	UGC	Cysteine, cys
	UUA	Leucine, leu	UCA	Serine, ser	UAA	Stop	UGA	Stop
	UUG	Leucine, leu	UCG	Serine, ser	UAG	Stop	UGG	Tryptophan, try
C	CUU	Leucine, leu	CCU	Proline, pro	CAU	Histidine, his	CGU	Arginine, arg
	CUC	Leucine, leu	CCC	Proline, pro	CAC	Histidine, his	CGC	Arginine, arg
	CUA	Leucine, leu	CCA	Proline, pro	CAA	Glutamine, gln	CGA	Arginine, arg
	CUG	Leucine, leu	CCG	Proline, pro	CAG	Glutamine, gln	CGG	Arginine, arg
A	AUU	Isoleucine, Ile	ACU	Threonine, thr	AAU	Asparagine, asn	AGU	Serine, Ser
	AUC	Isoleucine, Ile	ACC	Threonine, thr	AAC	Asparagine, asn	AGC	Serine, Ser
	AUA	Isoleucine, Ile	ACA	Threonine, thr	AAA	Lysine, lys	AGA	Arginine, arg
	AUG	Start, methionine, met	ACG	Threonine, thr	AAG	Lysine, lys	AGG	Arginine, arg
G	GUU	Valine, val	GCU	Alanine, ala	GAU	Aspartic acid, asp	GGU	Glycine
	GUC	Valine, val	GCC	Alanine, ala	GAC	Aspartic acid, asp	GGC	Glycine
	GUA	Valine, val	GCA	Alanine, ala	GAA	Glutamic acid, glu	GGA	Glycine
	GUG	Valine, val	GGG	Alanine, ala	GAG	Glutamic acid, glu	GGG	Glycine

Let's see a brief example of how this works, using a picture:

8 There are technically a few variations of this table depending on the organism, but most living systems use this one. Also, this table is listed in RNA form. Remember that the U's are T's in DNA. So, for example, encountering the DNA nucleotide triplet of TTG means leucine, ACT means threonine, and so forth.

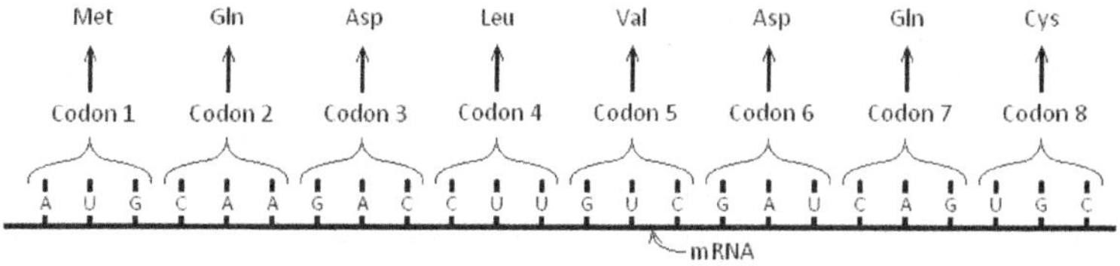

In this image, you can see that AUG causes a methionine to be placed in the first position. Then CAA causes glutamine to be placed next, and so forth. In this manner, the codons determine the order of the amino acids in the protein once the codes of the codons are translated.

Notice from the genetic code that the codon AUG represents both a start signal and methionine. Also, there are three codons that represent termination signals. These tell the cell when the end of the amino acid order has been reached. Do not confuse these "start" and "stop" codons with the start and stop signals we saw during RNA transcription. They are different processes. Before we can explore that further, however, we'll need to go over the details of transfer RNA.

Transfer RNA

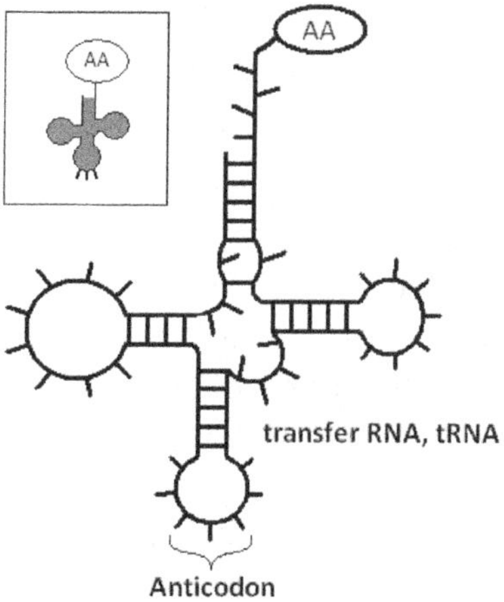

We spoke briefly about tRNA during the last chapter. These molecules bridge the gap between the codons of the mRNA and amino acids, and their structure is critical in the process. Transfer RNAs are single-stranded chains of RNA nucleotides that have coiled up to form a tangled blob roughly shaped like a bent "t."

At the top is the amino acid. At the bottom is the **anticodon**, which consists of three RNA bases pointed down like a diverging pitchfork (previous image). During the course of RNA translation, the tRNA matches up its anticodon with the codon on the mRNA, and in doing so marks that spot for the amino acid. Once in place, another tRNA bearing the next amino acid lands on the next codon, as shown below.

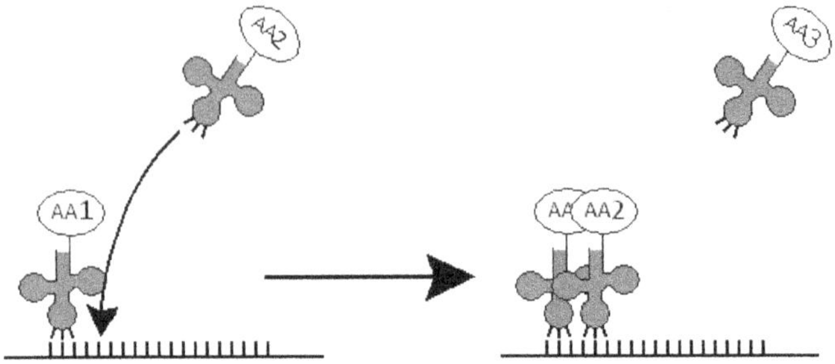

So how many different tRNA's exist? Well, the answer depends on the species. Let's take a guess though. A quick look at the genetic code will tell you that there are 64 total codons (UUU, UUC, UUA, UUG . . . GGU, GGC, GGA, GGG). So our first guess is that 64 tRNA are necessary. Not quite. We have to subtract 3 because three of the codons are translated as termination codons; they don't code for amino acids. We're down to 61, which is a good guess. As it turns out, however, there are *less* than 61 types of tRNA in most living systems. Why?

The reason there are less than 61 tRNA is due to a process described by the **Wobble Hypothesis**. If you take another look at the picture of the tRNA, you'll notice that the anticodon is curved. This means that the first two bases can match up with the first two bases of the codon, but the last one has poor contact and contributes less distinction to the matching process.

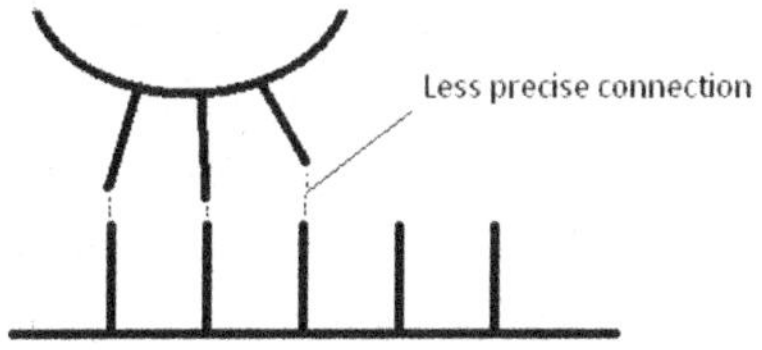

As a result, one anticodon might recognize up to four different codons, depending on the identity of the third base. This is why several codons code for the same amino acid in the genetic code. Because of this principle, the actual number of required tRNA drops to approximately 31, depending on the species. Some require more, and some require less, but 31 is an average number.

We are almost done with our preliminary discussion and will soon be looking at RNA translation. There is, however, one more question we need to ask: How do the amino acids get matched up with the right tRNA? After all, if proline, say, does not get correctly matched with the tRNA that has the anticodon GGG, it will not get correctly placed when the codon CCC appears on the mRNA. The job of making sure that the right amino acid is attached to the right tRNA comes down to a group of proteins called **aminoacyl tRNA synthetases** (think of this as amino acid-onto-transfer-RNA synthesis enzyme). With the help of these enzymes, the cell has a whole collection of tRNA molecules, each with their matching amino acids, just waiting for RNA translation to begin.

Initiation of Translation

We are now ready to talk about the initiation of RNA translation. This process is like catching a fly and caterpillar in your hands using bug bait (see image on the following page). If you recall, the two halves of a ribosome look like little hands. The smaller half, (or small subunit), starts the translation process by binding small proteins called **initiation factors** (i.e., the bug bait). Once these factors bind, the small unit is then ready to attract the mRNA (the caterpillar) as well as the first transfer RNA, (the fly). This tRNA is carrying methionine. The mRNA attaches at a very specific place on the small subunit, matching up a sequence of its bases called the **Shine-Dalgarno sequence**[9] with a spot on the subunit called the **anti-Shine-Dalgarno sequence**. This process stabilizes the mRNA. At almost the same time, a formylmethionine-carrying tRNA (f-Met-tRNA) matches its anti-codon with the AUG codon on the mRNA, which appears shortly after the Shine-Delgarno sequence[10].

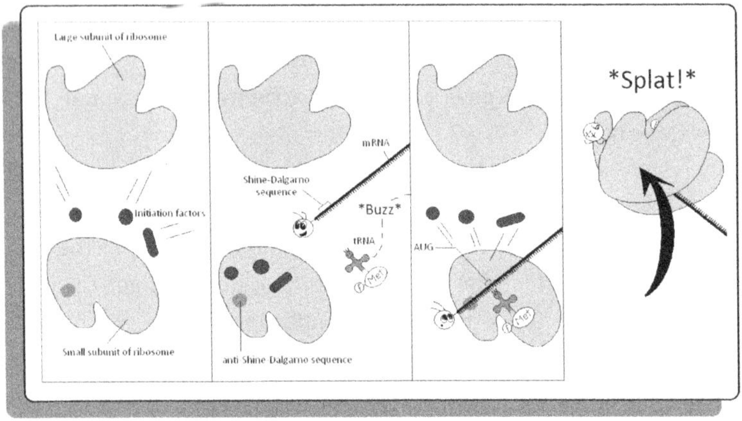

Once the f-Met-tRNA has bound to the starting AUG codon (it is Met because AUG codes for Met) and the messenger RNA is firmly attached to the Shine-Delgarno area of the small subunit, the large subunit attaches, much like two hands coming together to squish the fly and caterpillar in between. RNA translation is now ready to begin.

Notice, however, that "AUG" codes for both "start" and for methionine. This, naturally, begs an important question: Do all proteins have to start with methionine? The answer, surprisingly, is no. The methionine on this first tRNA has a small "formyl" chemical

---

[9] Technically, the Shine-Dalgarno sequence is found in prokaryotes. In eukaryotes, it's called the Kozak sequence. The function is essentially the same in both.

[10] The Shine-Dalgarno sequence is six bases long, usually AGGAGG, and typically appears eight bases before the AUG.

group attached to it. This formyl[11] group is a signal to the organism that this methionine is to be removed before the protein becomes mature. Only the first Met-tRNA is marked this way. The other Met-tRNAs are normal.

The Translation Cycle

Once the mRNA and first tRNA are trapped between the two subunits, a series of repeating steps commences. For reference, see the image below. In that image, we pretend the small subunit is invisible so that we can observe the inner workings of this cycle, step by step. The mRNA is positioned horizontally across the large unit and is facing toward the left. The first tRNA is already in position, and in this case, we say it is resting in the **P-site**. At this point, the first thing to happen is that a tRNA matching the next codon enters and

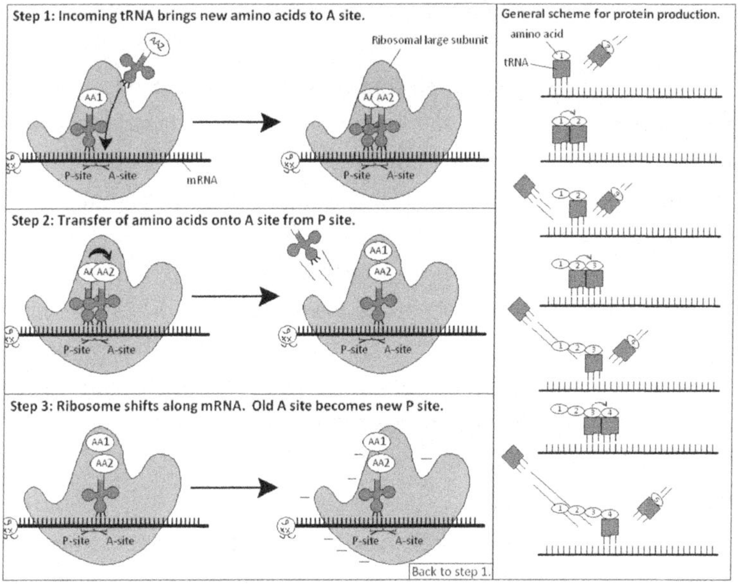

---

[11] A formyl group is just a HC=O attached via the C to another molecule.

attaches to the mRNA at the **A-site**. Once the new tRNA is bound to the A site, the old tRNA transfers its amino acid onto the amino acid of the new tRNA. The old tRNA then departs. Once this happens, the ribosome moves along the mRNA so that the remaining tRNA is shifted into the P-site, and the A-site is freed for the next tRNA to enter. The energy molecule GTP is used to move the ribosome and fuel the cycle. The steps then repeat and will continue to do so until a termination codon is reached.

To the right of the image is a schematic for the order in which amino acids enter into the reaction and how these amino acids become linked into a chain. Basically, each new tRNA that enters adopts the existing chain of amino acids and allows the previous tRNA to leave. It then waits until the next tRNA enters, whereupon it gives the amino acid chain (plus one amino acid) to the new tRNA, and so forth.

When one of the three stop codons is encountered, a **release factor** protein binds to it and causes the entire structure to come apart; the subunits detach from the messenger RNA and the new protein is freed. The release factor is sorta like my roommate; he shows up and all my guests bail.

If you'll recall, the large ribosomal subunit has a tunnel passing through it, like a hole in the palm of the hand. The growing chain of amino acids passes through this tunnel (not shown above), and emerges from the other side in the order in which the amino acids were added. Once they begin to emerge from this tunnel, their order determines the fate of the protein. Let's talk about that next.

Protein Destination

If you remember, prokaryotes (i.e. bacteria) have no nuclear membrane. Their mRNA also do not go through the post-translational changes, such as 5' capping, poly-A-tail addition, intron splicing and all that mumbo-jumbo. In fact, protein translation in prokaryotes can start before the mRNA is even done being transcribed from the DNA!

In electron microscope images, one can sometimes see ribosomes attached to mRNA emerging from the DNA. The new protein being translated is too small to see in these structures, however, so the whole thing just looks like little blobs on strings.

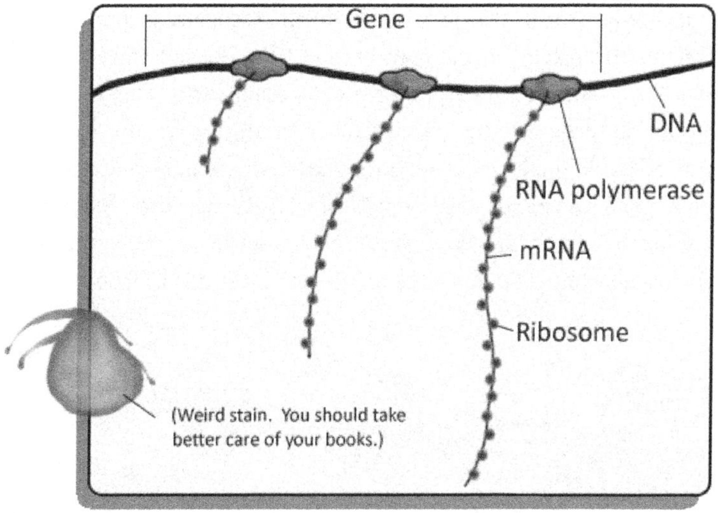

In eukaryotes (such as you and me and possible the odd thing growing in my sink), there is a cell nucleus that separates the events of transcription from translation. Once the pre-mRNA has gone through all the modifications and becomes a mature mRNA, it leaves the nucleus via a nuclear pore and is caught by a ribosome floating in the cytoplasm. For most mRNA, translation then proceeds (as described above) right there in the cytoplasm. The proteins are created, released and simply float away. For other mRNA, however, the process may be temporarily paused by a **signal recognition particle**[12].

[12] The signal recognition particle is not a particle like the protons, electrons, and neutrons. It's a protein/RNA complex. I normally refer to it as the signal recognition complex, but you'll see it written as the signal recognition particle in texts and publications, so I've presented it as such.

Basically, the signal recognition particle "watches" for a specific sequence of amino acids emerging from the tunnel of the large subunit. If the first few amino acids match this sequence, the recognition particle binds to them and causes translation to temporarily stop. The ribosome/mRNA/recognition particle then attaches to proteins in the membrane of the endoplasmic reticulum, where translation resumes, only now the amino acid chain is threaded through a pore in the membrane. In this way, the protein is synthesized directly into the inside of the endoplasmic reticulum. Since the ribosome and mRNA are stuck to the wall of the endoplasmic reticulum throughout the translation process, this makes the endoplasmic reticulum look rough with little specks stuck to it.

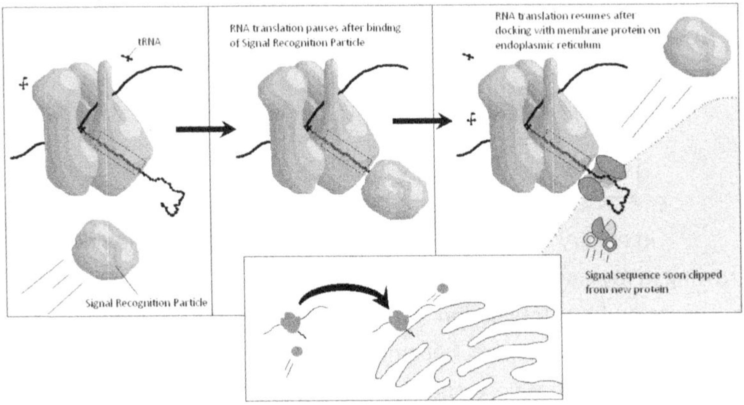

The proteins that are threaded into the lumen of the endoplasmic reticulum in this way are destined for secretion from the cell, or will be incorporated as membrane proteins (such as those in the cell's membrane), or will be routed to organelles such as lysosomes to be used in the cell's digestion process. The signal sequence that caused all of this to happen is clipped before the protein is fully mature.

Translation Time

All of these impressive steps take place rather quickly. For example, ribosomes can add between 6-21 amino acids to a growing chain per second, depending on the amino acid and the species in question (eukaryotes being slower). That seems fast. After all, there isn't much *anyone* can do *that* many times a second. As it turns out, however, that isn't fast enough to sustain life. So how do organisms get around this? Well, there are several mechanisms. First, multiple copies of mRNA are typically being made at any given moment. So if each one is being translated, that increases the total rate of protein production. Secondly, two or more ribosomes can often bind to the same mRNA. While one is finishing up, another could be in the middle of making protein and still a third could just be starting! So, if the cell has 100 mRNA in circulation, protein production speeds up by a multiple of 100. If two or three ribosomes can fit onto the mRNA at once, protein production speeds up by another multiple of two or three. So how may mRNA do cells usually have floating around? Typically, they have on the order of *several hundred thousand*. So, if the cell has 100,000 mRNA and two to three ribosomes attached to each one, it means that amino acids are being added to growing protein chains at a rate of roughly three to four million amino acids per second. Now, *that* is fast, but those numbers are rather conservative. Usually many more than two or three ribosomes can fit on a messenger RNA, increasing the total rate of amino acid addition even more.

Summary

In this section, we discussed the basics of RNA translation.

- RNA translation is the second half of the Central Dogma of Biology, where the mRNA is used to determine the order of amino acids to add to a new protein.
- Nucleotides triplets (3 nucleotides), called codons, are each assigned to an amino acid. AAA goes to lysine, GGA goes to glycine, CCU goes to proline, and so forth. Often, many codons are assigned to the same amino acid. For example, GGU, GGC, GGA, and GGG all are assigned to glycine. The complete list of codon assignments is called the genetic code.
- The order in which codons appear in the mRNA determine the order in which amino acids are added to a new protein.
- Transfer RNA, or tRNA, are 't' shaped single-stranded RNA blobs that have an anti-codon on one end and an amino acid on the other end. There are roughly 31 types of tRNA, on average, in most living systems. Each tRNA has its own anticodon, which pairs with a codon in the genetic code. Each tRNA is matched with the appropriate amino acid, a reaction that is controlled by enzymes called amino acyl tRNA synthetases.
- The Wobble Hypothesis predicts that some tRNA can fit onto more than one codon due to imprecise matching of the third base pair in the anticodon on the tRNA. This is due to the curved structure of the anticodon on the tRNA and is a big reason why multiple codons are assigned to single amino acids.
- Initiation of RNA translation starts when the small ribosomal subunit binds initiation factors. Once bound, mRNA and the first tRNA binds to the small subunit, the initiation factors

leave and the large subunit attaches to the complex. The mRNA/ribosome is now ready to begin.

- RNA translation proceeds in a repeating cycle of steps.
- A new tRNA/amino acid enters and attaches at the A-site, as dictated by the codon that appears there.
    - 1) The amino acid chain is transferred to the new amino acid, and the old tRNA at the P-site leaves.
    - 2) The ribosomes moves along the mRNA so that the new tRNA is shifted from the A-site to the P-site. The process now repeats until a stop signal is found.
- When a termination codon appears at the A-site, a termination factor protein binds and causes the ribosome/mRNA/amino acid complex to come apart. The new protein is freed.
- During construction, the new protein is threaded through a tunnel that appears in the large subunit of the ribosome. As the new protein emerges from the other end of the tunnel, a signal recognition particle watches it for a specific sequence of amino acids. If the specific sequence is found, the signal recognition particle grabs the new protein end and causes production to pause while the entire thing moves and attaches to a channel protein in the membrane of the endoplasmic reticulum. When this happens, the new protein is routed through the channel into the space inside the endoplasmic reticulum. New proteins routed to the endoplasmic reticulum are destined to be either 1) secreted from the cell, 2) embedded into one of the many membranes of the cell, or 3) become part of the vesicle system, such as in lysosomes.

- Cells typically have hundreds of thousands of mRNA copies floating around, and multiple ribosomes can embark on RNA translation at the same time. This drastically increases the rate at which proteins can be produced.

Unit 4

Viruses and Transposons

In this section we cover:

1) Viruses

2) Transposons

In previous sections, we covered how DNA is reproduced. Then we switched tracks momentarily to talk about how the cell works hard to make proteins in the Central Dogma of Biology. Now, let's switch tracks again to talk about another key topic. For every hard worker, there is a moocher who lives off of him (or her). Politicians, bill-collectors, useless roommates – it appears there is no end. The same is true in biology. In the cellular world, there are entities whose sole purpose is to hijack cells and use their machinery to replicate and make their own protein. The most prevalent example is the **virus**. To this list, one might add a second entity known as the **transposon**. Let's discuss them one at a time.

Viruses

Viruses are packets of DNA or RNA encircled by proteins that invade a target cell and then take over the protein-making machinery of its cellular victim. In essence, they supply the genetic information while the cell supplies everything else. The virus' genetic information is replicated, and their proteins are made. New viruses are then put together from these new parts, and once assembled they escape from the cell to repeat the cycle. Often, the cell that is invaded is killed when the viruses leave.

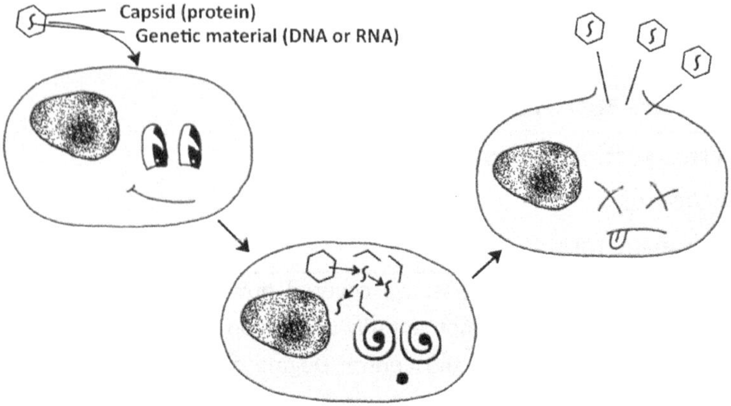

Viruses have relatively simple structures. Their shell, called a **capsid**, is generally made from repeating units of proteins. The capsid is usually one of two different shapes, tubular or icosahedral, although more complex shapes also occur. Inside the capsid is the genetic information, (which is either DNA or RNA). Around the outside of the capsid, some viruses have a lipid envelope. This envelope is the remnant of the lipid membrane from the last cell victim.

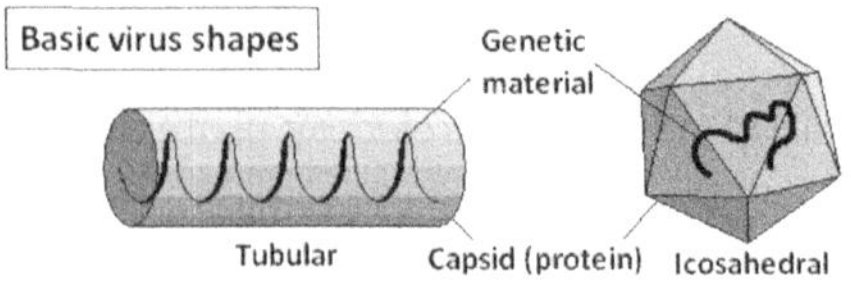

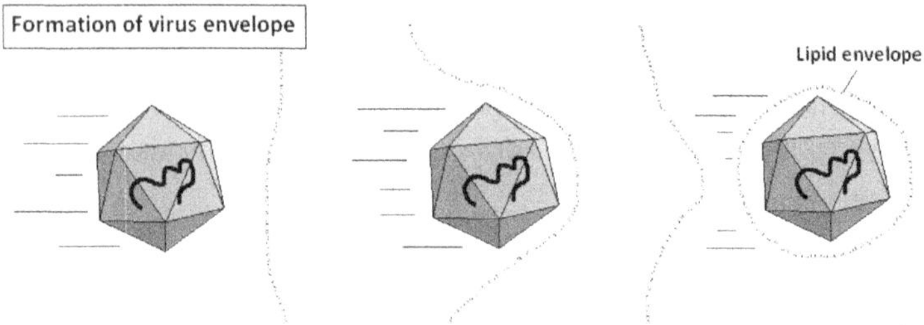

Apart from shape, there are two broad types of viruses: DNA viruses and RNA viruses. Each of these can be divided into single-stranded (ss) versus double-stranded (ds), referring to whether or not the genetic material comes packaged as just one strand or as a helix. The shorthand is to apply "ss" or "ds" in front of the virus to indicate the strandedness. For example, "ss-RNA virus" means a single-stranded virus that has RNA for its genetic information. As a rule of thumb, most DNA viruses are double-stranded, and most RNA viruses are single-stranded.

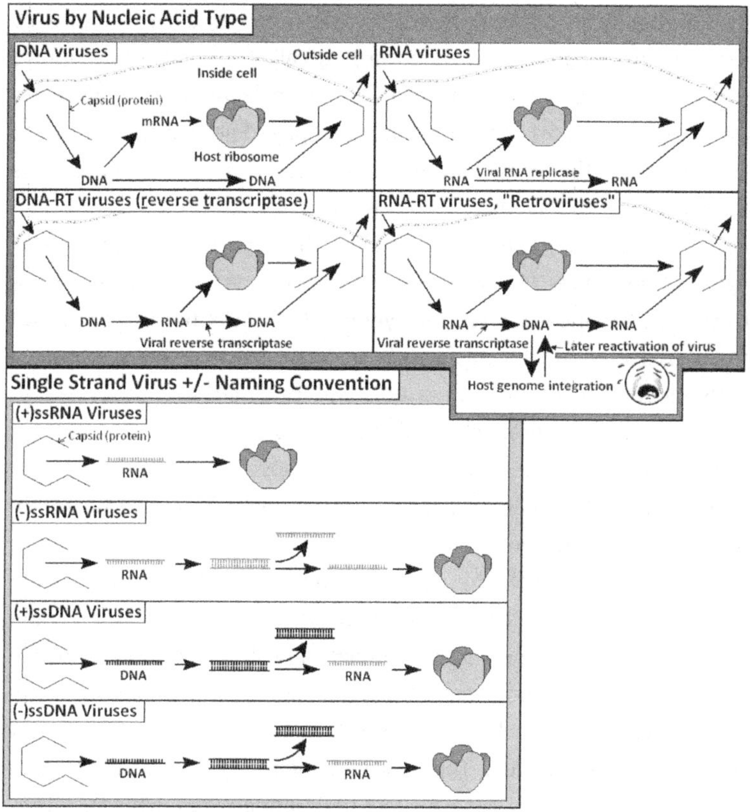

Further, single-stranded viruses are given a (+) or (–) designation to signify whether or not they are similar to or complementary to the equivalent mRNA that makes its proteins. For example, if a single-stranded RNA virus busts into a cell, and its genetic strand heads directly to the nearest ribosome, we would say that the strand is equivalent; the virus is called a (+)ss-RNA virus. If, on the other hand, the virus first has to make a complementary strand using a polymerase, and then that newly made strand goes to the ribosome, we would say the virus is (–)ss-RNA. The distinction between the (+) and (–) types is summarized in the diagram on the previous page.

Lastly, as if this were not complex enough, each type of virus (DNA vs RNA) has a subclass of viruses called reverse transcriptase viruses. These viruses make DNA from RNA (the reverse of transcription) as part of their replication cycle. Some of these kinds of viruses are further classified as retroviruses if that resulting DNA is then inserted into the host DNA. Their genetic information can then be reactivated sometime later to start the replication cycle all over again. These kinds of viruses are arguably the nastiest, because once their DNA has incorporated into yours, it's REALLY hard to get rid of them. HIV is known for this, which is why infection with it is – at least with our current technology – a life-sentence of illness.

It's not entirely clear where viruses originated. The current belief is that viruses have one of three possible origins. The first hypothesis is that they may have originally been primitive **bacterial-like parasites** that burrowed into cells. Over the millions and millions of years of life on Earth, these parasites became so dependent on their hosts that they lost their ability to make their own ribosomes and other organelles. They became simpler in design, until only a protein capsule and genetic information was needed in order to hop from one cell to another. Mitochondria and chloroplasts are examples of how this may have happened. These two organelles have a LOT of things in common with bacteria but have lost their independence because they can no longer make the things that their host cells supply to them. Perhaps in the distant future, mitochondria and chloroplasts will become simpler still, until one day they would hardly look like bacteria at all. Who knows?

The second hypothesis is that viruses coevolved with cells. Although we will be delving into this topic of evolution much more at length in Topic Text D, I will discuss it here in passing. The most primitive form of life is believed to have used RNA as its genetic information (long before DNA). In those times, it is believed that viruses and cells probably didn't look much different from each other. Like my roommate, one lived off the other, and the relationship continues today. The viruses didn't go far in life, while the cellular organisms basically took off for great complexity.

The third hypothesis is that viruses started off similar to entities that we observe in labs every day. These entities are called **transposons**, or "jumping-genes." They have greater relevance than just their *hypothetical* start as viruses, however, so we'll devote an entire section to them.

Transposons

Transposons are sequences of DNA that have the ability to be cut out of one section of DNA and reinserted into another. It's like a person who cuts in line at random, then steps out of the line to pick a new spot somewhere else. These stretches of DNA are bordered by short sequences at both ends, and these ends are recognized by a protein called **transposase** that cuts the sequence out (putting the cut ends back together) and moves it somewhere else in the genome.

Many transposons contain the gene for transposase, making them self-actuating jumping genes. Other transposons do not, and they jump from one location to another simply because *other* transposons with the transposase are in the *same* genome. Still other transposons do not jump at all, but rather are *copied* and inserted into the new location. When the event is done the transposon will exist at both the original *and* the new location.

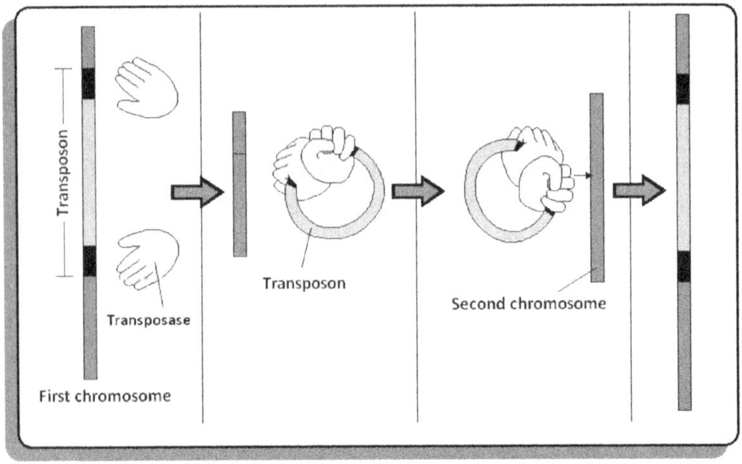

At any rate, the similarity between transposons and viruses is uncanny: they are both stretches of genetic information that contain the blueprints for proteins that allow them to hop from one host to another. The theory is that viruses slowly evolved from these entities. The first form of viruses was little more than a simple transposon. Over millions of years, transposons slowly became more adept at

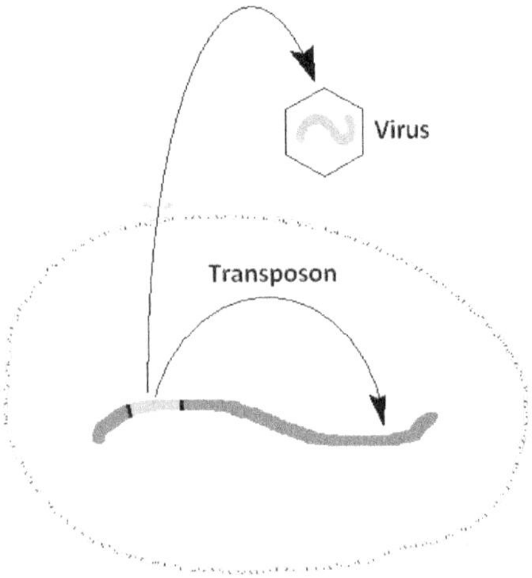

jumping from chromosome to chromosome until they finally developed the ability to go from one cell to another. It's a lot like mankind's progression to the stars. First we had to overcome our fear of gravity. Then we developed balloons. Then we developed planes, then jets, then space rockets, and someday (hopefully soon) we will have bases on Mars. It's all about baby steps. Primitive man was like transposons. We jumped from one place on Earth to another. Modern man is becoming a virus, packaging little parts of Earth together so we can infect another planet.

Granted, at this stage it may not seem entirely clear just how a transposon gains the ability to go from jumping between chromosomes to jumping between cells. After all, where did these extra genes come from if they weren't made in a lab somewhere? Well, before we go any further, I must posit a warning: the belief that viruses were originally transposons is just a theory. That said – starting in the next section we are going to begin exploring where new genes come from, and hopefully then these ideas will be a little clearer.

Summary

In this section, we discussed viruses and transposons.

- Viruses are mobile packets of genetic information that invade cells and utilize their protein-making machinery to complete each viral replication cycle. They do not have ribosomes or any other organelles.
- Viruses are comprised of a protein capsule, known as a capsid, genetic material inside the capsid, and sometimes a lipid envelope surrounding the entire thing. The genetic information is either DNA or RNA. The lipid envelope is the remnant of the last victim's plasma membrane.
- Both RNA and DNA viruses can be single stranded "ss-" or double stranded "ds-". Most RNA viruses are single stranded. Most DNA viruses are double stranded.
- Single stranded viruses have a designation of either (+) or (–). Positive stranded viruses are similar to the equivalent mRNA that codes for the protein. Negative stranded viruses are complementary to the mRNA that codes for the protein.

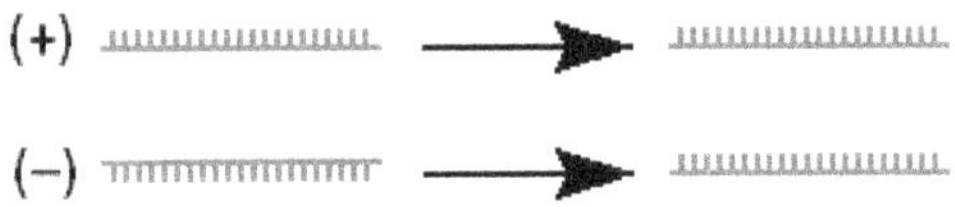

- Both DNA and RNA viruses have a subset of viruses called reverse transcriptase viruses, suffixed "-RT." These viruses use a protein called reverse transcriptase to make DNA from RNA. This is unusual in that it is the reverse of what cells do when they make protein. An even smaller subset of RNA-RT viruses are known as retroviruses, because the DNA they make gets inserted into the host's genome for later reactivation. HIV is an example of such a virus.
- Viruses may have one of three proposed origins. 1) Regressive evolution from bacteria-like parasites. Over time,

they lost their ability to make their own proteins and have stripped down to the bare essentials in order to go from one cell to the next. 2) Co-evolution with cells since the first days of life. Cells evolved into cells, whereas viruses didn't advance much at all. 3) Viruses may have started off as transposons and gradually evolved to become more complex over time.

- Transposons are segments of DNA that are capable of being cut from one area of DNA and inserting into another. They usually are flanked by distinct regions of DNA recognized by an enzyme known as transposase that is usually encoded by the transposon's genetic information. Transposase cuts the transposon from where it is and moves it somewhere else. Some transposons work by being copied instead of cut, so that over time they increase in number and location.

Unit 5

Mutations

In this section we cover:

1) What mutations are.
2) The effects of mutation.
3) The causes of mutation.
4) What mutation do *not* do.

We have talked at great length about how the genetic code is used to make proteins, but a lingering question remains: how did the code get into that sequence? No, it did not happen because of a scientist in a lab. The scientist is only there because his genetic sequence allowed him to live and breathe and work and so forth. Therefore, we can't blame genetic sequence on mad scientists, because it puts the cart before the horse – so to speak. To help answer this question, we will now talk about the most important mechanism that drives the formation of genetic sequences: **mutations**.

Mutations

Mutations are changes in genetic sequence. For example, a change from ATTCG to ATCCG represents a type of mutation. In this case, the mutation is called a **point mutation**, because the change was only at one place in the sequence, here T to C. Other types of mutations exist as well. For example, an **insertion mutation** inserts one or more nucleotide bases into the sequence. So, ATTCG might become ATT[CGAGACCT]CG where the bracketed part represents the nucleotides inserted. Yet another type, **deletion mutations**, does the opposite: it omits X number of nucleotides from a sequence. ATTCG might become A[]G, where the brackets represent the missing nucleotides. Each of these types has numerous causes, some of which we will explore later in this chapter. For the moment, let's focus on the *consequences* of these changes.

If a mutation occurs inside a gene – or other area of DNA serving an auxiliary role – then a serious error could result. The protein that a gene makes might have a different amino acid at some point, or a completely different sequence altogether, causing the protein to behave differently. Almost always, the change causes the protein to behave less well, giving the organism a disadvantage in life. It may even cause the organism to die prematurely due to internal chemical problems. A whole host of genetic diseases fall under this category: cystic fibrosis, sickle cell anemia, Huntington's disease, and so forth, to name a few.

Rarely, the mutation might cause the protein to behave more efficiently, in which case the organism has an advantage in life. Such an individual will tend to survive to mate more often than not. Each time it mates, the gene can be spread to more and more offspring. Over generations, the new beneficial gene becomes more frequent until the population at large has been "upgraded" with the good mutation. It needs to be double-stated for emphasis, however, that mutations rarely cause good things to happen. In short, you do not want a mutation. The odds are *very* unlikely that it will benefit you.

It may help to consider the following analogy: Imagine that you open a lengthy and well written report that is due in one hour. With eyes closed, you randomly click somewhere in the document and then randomly hit one or more letters on the keyboard. Finally you hit save. What are the odds that this behavior will result in a better report? Conversely, what are the odds that this change will just look like junk in the middle of a well written document? If you repeat this exercise a million times, ONE of those times may result in an improvement to the paper. Otherwise, you probably just dropped the quality of the report. Mutations work the same way.

That being said – let's zoom in on the actual mechanisms by which mutations cause problems. For point mutations, the alteration in genetic code usually results only in the change of a single amino acid, or an **amino acid substitution**, as it is called. This happens because the codons of the mRNA change, so when the ribosome reads the codon it matches a different amino acid than what was originally in that place. Sometimes, this change can insert a premature stop codon instead of an amino acid. In that case, the result is usually disastrous, because the protein is incomplete and totally useless. The image below shows these two possibilities. The first is an amino acid substitution that switches glutamine (gln) with glutamic acid (glu). The second introduces a premature stop codon.

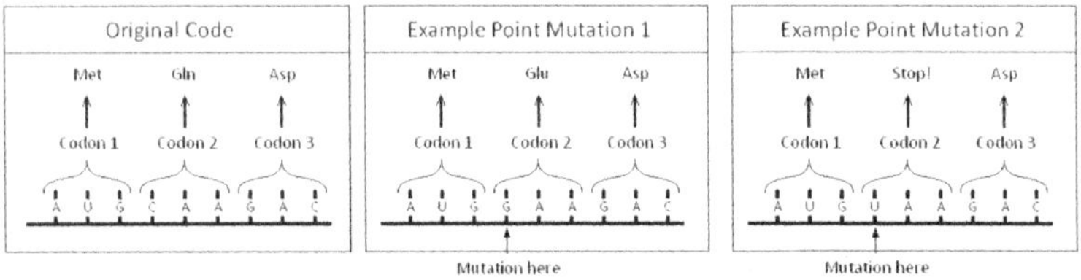

Aside from premature stop codons, point mutations are usually the least alarming of all the types of mutations. For example, if an amino acid substitution switches two amino acids that are of the same class (both acidic, or basic, or hydrophobic, etc.), the effect can often be negligible. If so, the mutation is considered silent, since it doesn't alter the behavior of the enzyme or cause any trouble to the organism. If the substitution occurs at some pivotally important part of the protein, however, the effect of the change could be disastrous. The mutation might shut the protein's activity down altogether, usually resulting in hardship for the organism having the mutation. A common example of this kind of mutation is the one leading to **sickle cell anemia**, where a protein in red blood cells called hemoglobin has a single mutation that switches one of its glutamic acids into valine. This causes the protein to change abnormally under certain physical conditions. When this happens, the red blood cells can get stuck in the small blood vessels of the body, causing that part of the body to begin to starve of oxygen. This simple mutation leads to a lifetime of incredibly deadly, painful crisis and frequent hospital visits. One last comment should be made about point mutations before moving on to other mechanisms. If the point mutation occurs at the third nucleotide of a codon, the change can often have no effect on the amino acid sequence whatsoever. This is because of the mechanism described by the Wobble Hypothesis. For example, a codon change from CCC to CCA will have no effect, since both sequences code for proline.

Insertion and deletion mutations suffer from the same problems as point mutations in that they alter the codons and thereby introduce the possibility of amino acid substitutions and premature stops. They also introduce another particularly nasty problem called **frameshift errors**. A frameshift error is when a number of nucleotides indivisible by three are introduced into (or removed from) a sequence of codons. When this happens, the nucleotides shift over one or more places, so that the codons are no longer in the correct groups of three as they once were. From that point on, all the codons are read out of frame and each one produces a *completely* different sequence of amino acids. Here is an image to illustrate the concept:

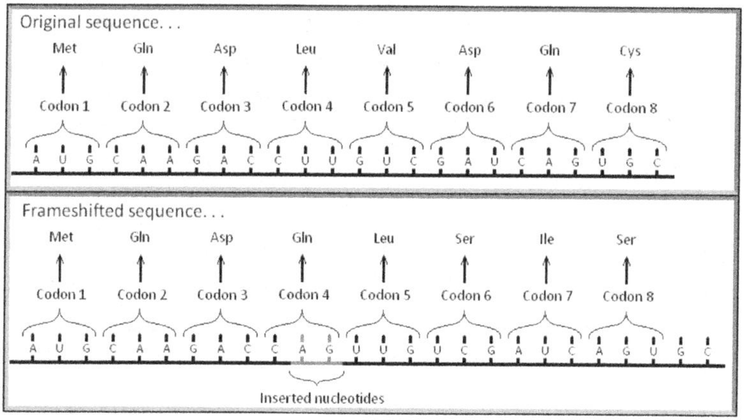

In this image, we see that the first three codons of each sequence code for met, gln, and asp, in that order. After that, the frameshift mutation in the second sequence causes all subsequent codons to be read incorrectly, resulting in a completely different protein.

Finally, there is one last mechanism in which mutations can cause changes in an organism. As we saw earlier, not all of the genome is used to code for protein. Apart from the vast amounts of junk DNA, there are also sequences that are used to regulate the rate of transcription of various genes. In other words, these rate controlling sequences control how often some other proteins are made. Let's take an example. Suppose we have a rate controlling sequence that governs how quickly "gene X" is used to make its protein, "protein X." Suppose, for the sake of argument, that the sequence increases the frequency of gene X's transcription. If a mutation occurs in the rate controlling sequence, it will likely interfere with the sequence's job. As a result, gene X will be read less often, and protein X will drop in concentration. The delicate biological balance inside the cell will be destabilized. The cell will become sick and probably die. In the reverse scenario, let us suppose that the purpose of the rate controlling sequence is to slow down the frequency of gene X's transcription. In that case, a mutation in the rate controlling sequence will cause protein X to be made more often, and its concentration will increase. Like before, this can destabilize the cell and lead to sickness and/or death.

To summarize, mutations can lead to changes in a protein's structure or rate of production. In the vast majority of cases, these changes are hazardous to the health of the organism, because the organism depends on a delicate chemical balance in order to exist. Having investigated how mutations result in biological changes, let us now look at what makes mutations occur.

Causes of Mutation

One important source of mutations occurs during DNA replication. The DNA polymerase machinery is not perfect. Sometimes, for example, the DNA polymerase slips a little off the strand it is reading, and when it regains its footing it might not be exactly where it left off. This can lead to gaps or duplication in the new strand. Sometimes a stray H^+ will slip between the base pairs, changing the way they pair up so that DNA polymerase inserts the wrong partner on the new strand. These are just two possibilities. Though there is some proof-reading capability in the polymerase enzymes, the rate of error has been estimated to be somewhere in the vicinity of one in a million base pairs. Human cells have just over 3 billion base pairs, which means each time one of our cells replicates, it experiences about 3,000 mutations! The rapidly dividing organs, such as the lining of the GI tract and blood cells, experience the most mutations, simply because they divide the most. Yep. Just by sitting there, you're mutating. Over the years, these mutations slowly accumulate, and sooner or later they occur in a place that screws up the control of the cell cycle. Once this happens, the cell will begin replicating out of control, becoming a cancer.

Another source of mutations occurs from mutagenic chemicals. While there are huge varieties of such chemicals, several general patterns can be observed. For instance, some are carbon-ring structures that hide in between the ladder steps of the DNA helix, (a process called **intercalation**). By doing so, they interfere with normal binding or appear as just another DNA base. When replication occurs, the polymerase misreads this invading chemical as a base and tries matching it with the "appropriate" base pair. This naturally changes the base sequence.

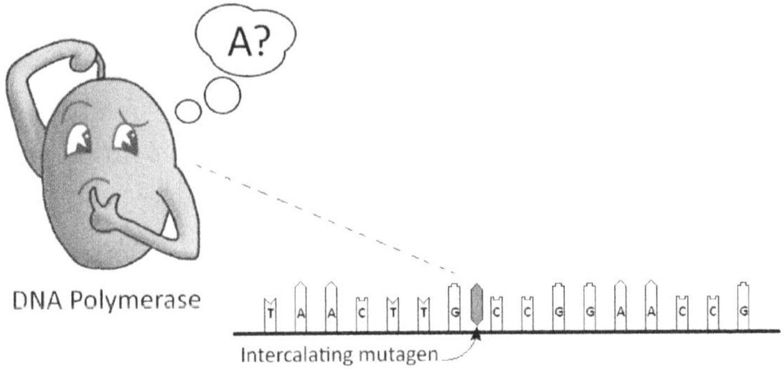

Alternately, some mutagenic compounds are structurally close enough to nucleotides that they are inserted into DNA at the time of replication. Later, these nucleotides shift their structure so that they look like a *different* base. Once the DNA replicates again, it is matched with the wrong base, causing a mutation. One example is 5-bromouracil, shown here:

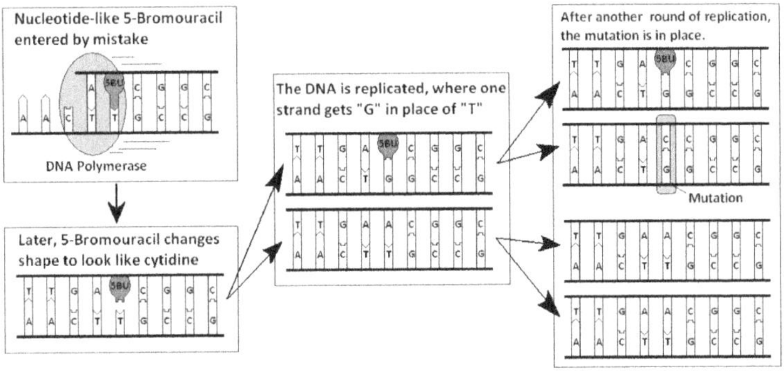

Other culprits of mutation that you'll likely hear about in popular media are **oxygen radicals**, which cause **oxidative damage** to DNA (and other parts of the cell). The more oxygen that is around, the more it tries to interrupt what DNA is doing (see unit 8, Topic Text B). It reacts with it, changing its structure. This, of course, causes all kinds of chaos for the cell. Cells do their best to repair the damage, but once in a while the damage causes a permanent mutation. Examples of common species capable of producing oxygen radicals include molecular oxygen (O_2), hydrogen peroxide (H_2O_2), superoxide anions (O_2^{2-}), and organic peroxide[13] (ROOH, where R is some arbitrary carbon structure). Some of these radical sources are found in commercial products. Others are just floating around in the air looking for trouble. The first example, O_2, gives us a love-hate relationship with this molecule. Without it, we die. With *too much* of it, or with uncontrolled presence in the body, we get sick. This is a big reason why plants do not horde O_2 after they make it.

Having explored the topic of mutagens, let's move on to the other culprits in the mutation-causing business. The next on the list is radiation, which can take many forms. Radiation is a general term that refers to both electromagnetic radiation and nuclear decay. *Electromagnetic* radiation is the large spectrum of light particles traveling in waves. We discussed these briefly in the sections on photosynthesis in the previous topic text. The light that we see is part of that spectrum. *Nuclear decay* was discussed briefly near the beginning of the first topic text when we talked about atoms. It refers to the various pieces that spring out of the unstable nuclei that are breaking down. Think of nuclear decay as a stream of "heavy" particles like protons, neutrons, and (oddly enough) electrons,[14] whereas electromagnetic radiation is a stream of light particles.

Either way, both types of radiation can be harmful to DNA. Ultraviolet light, for example, is a form of electromagnetic radiation that causes the nucleotides to rearrange internally, forming a localized

---

[13] Regular peroxide, HOOH, will also cause oxidative damage.

[14] It may seem odd that electrons can fly out of the nucleus, especially since they don't hang out there to begin with. In nuclear decay, however, sometimes neutrons will split in half, producing the positively charge proton and the negatively charged electron. This electron zips off into the distance. Also, protons, neutron, and electrons are not the only entities generated by unstable nuclei. Some electromagnetic radiation is emitted from the nuclei. Indeed, this is the reason the sun emits so much light; its atoms are undergoing nuclear chemistry.

abnormality in the strand. Sometimes the cell's repair mechanisms can recognize this damage, but sometimes not. When it goes unrepaired, (or *incorrectly* repaired), a mutation results.

Stronger forms of radiation, like X-rays, gamma rays and nuclear decay, will often splinter the DNA backbone or break off nucleotides completely. Think of these as bullets shot at a rope ladder, tearing it all to pieces. As with UV damage, cells have repair proteins that search for characteristic damage caused by these kinds of radiation, but the repairs aren't always perfect. Once in a while a mutation sneaks in.

As if mutagens and radiation weren't enough, there are also biological causes of mutations as well. The first we'll discuss is due to transposons – those "jumping genes" we discussed in the last chapter. When these genetic drifters hop into DNA, they occasionally jump right into the middle of a gene, interfering with its transcription. Naturally, this changes the protein that is encoded there. Sometimes they insert into the sequence that control the rate of a gene's transcription, speeding up or slowing down the creation of the respective protein as we discussed above.

Another biological cause of mutation is due to viruses, or more specifically *retroviruses*. If you'll recall, retroviruses are RNA viruses that create a DNA template of their information and insert it into the genome of the host. When they do this, it can cause the very same havoc as transposons; they can plop their DNA right in the middle of your important DNA, completely messing up the code that is there, having the same consequences discussed previously.

Ending it all!

If exposed to too many mutations, regardless of the cause, the cell may begin to function abnormally. In response, these damaged cells may often undergo a deliberate process of suicide, called **apoptosis**. While this may sound counterproductive, the reason for the strategy is quite good. If a cell sustains too many mutations, the chances of uncontrolled cell division (i.e. cancer) increase. Apoptosis, therefore, is a survival mechanism put in place to protect the organism from the cancer-causing effects of mutation.[15] Unfortunately, this defense mechanism is not perfect either, because apoptosis is a process that requires proteins. If a mutation disrupts proteins needed for apoptosis, then the cell will be unable to perform suicide. In other words, if a cancer forms in that cell, it will not be able to kill itself, and the organism is in major trouble.

---

[15] There are other reasons and instances of apoptosis, however, but this is an important one.

Summary

In this section, we discussed mutations.

- Mutations are changes in the sequence of nucleotides in an organism's genetic material (usually DNA).
- There are three broad categories of mutations: 1) point mutations, 2) insertion mutations and 3) deletion mutations. Point mutations are changes in a single nucleotide. Insertion mutations are when one or more nucleotides are introduced into a sequence. Deletion mutations are when one or more nucleotides are cut from a sequence.
- Mutations alter the functioning of a living system by changing the codon of the messenger RNA or altering the rate of protein transcription. If a mutation occurs at the 3rd position of a codon, it may not always result in a change of amino acid type.
- Mutations are considered silent if they result in no change of biological function. Mutations occurring in junk DNA or point mutations in the 3rd location of codons are usually silent.
- If a mutation occurs within a stretch of DNA that controls the rate of transcription of some gene, the mutation may cause the gene to be transcribed at a new rate. It may speed the rate of transcription, resulting in more proteins being made, or it may slow transcription, resulting in less proteins over time.
- The causes of mutations include internal errors of DNA replication and repair, mutagens, radiation, and biological entities such as viruses and transposons.
- Mutagens work by one of several mechanisms: They can 1) intercalate between neighboring base pairs (thereby looking like a base), 2) be structurally similar enough to a nucleotide that they are inserted at the time of replication (and later

change shape to look like a different base), or 3) they can directly react with a base to change its structure.

- Radiation is a general term that refers to electromagnetic waves (the light we see being an example) or the particles that are emitted from unstable atomic nuclei. Both types interact with DNA, changing its structure or breaking its backbone. Repair mechanisms are not always successful at reversing the damage, and in those cases a mutation remains.
- Retroviruses and transposons can insert themselves in DNA. This behavior can result in mutations. If insertion takes place within a gene or rate controlling sequence, the chemical function of the cell will change.
- Mutations are almost always bad. On rare occasion, mutations can result in an improvement to the existing protein design. The rare improvements, likely, are rarely groundbreaking. Over the course of millions and millions of years, the result is a gradual trend toward steadily more efficient genetic designs.

Unit 6

The Cell Cycle

In this section we cover:

1) The Cell Cycle
2) Interphase (G_1, S, G_2)
3) Mitosis (Prophase, Metaphase, Anaphase, Telophase)
4) Cytokinesis
5) The Cellular Clock

The Cell's Day

In previous sections, we covered how DNA is reproduced. Then we switched tracks momentarily to talk about how protein is manufactured in the Central Dogma of Biology. We also saw how mutations creep into that scene and how viruses take advantage of the entire thing. Now let's have a look at a cell's typical "day." The cell spends its time alternating between two different states: **interphase** and **mitosis**. During interphase, the cell conducts normal business; it makes various proteins and spends at least part of its time replicating its DNA. During mitosis, the replicated DNA molecules are separated from each other, usually into two new cells. At the end of mitosis, interphase resumes and the process repeats. This entire cycle is known as the **cell cycle**, shown in the next image.

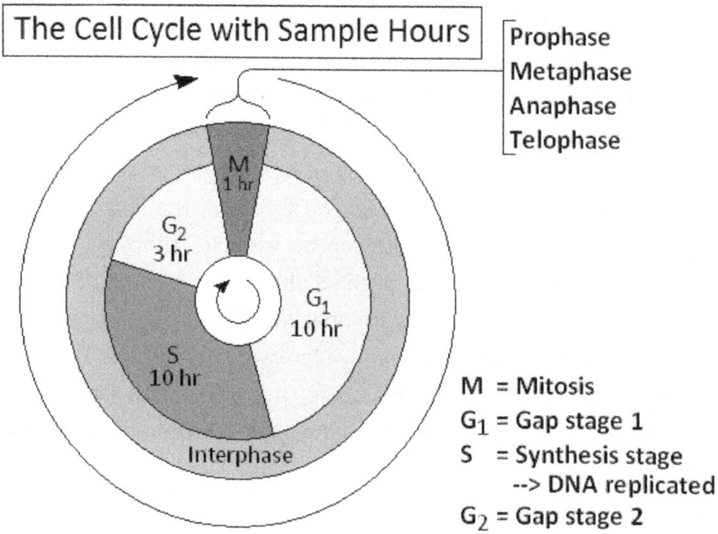

As you can see in the image, interphase is divided into three distinct stages: G_1, S, and G_2. Although the image assigns a certain amount of time to these individual stages, the hours given – though typical – are only samples. The actual time spent in the individual stages is variable and depends on the cell in question.

In the **G_1 stage**, (the "G" standing for "gap"), the cell begins replicating various organelles such as mitochondria, chloroplasts and ribosomes in preparation for mitosis. The length of time that a cell spends in G_1 is extremely variable. At one end of the spectrum are cells that are incredibly efficient at cell division, such as embryonic cells; they spend so little time in G_1 stage that they seem to steamroll right past into S phase. At the other end of the spectrum are cells such as muscle and nerves that *never* leave G_1. They never divide again. This means that the nerve cells you have at birth are the same you carry the rest of your life. They don't divide to make more, so take care of them or you'll end up like my housemate. It's a terrifying fate, because I've never met anyone else who could come up with an infinite supply of incredibly terrible ideas. It's like the time he convinced me to write biology books. I'm still paying for that one.

At any rate, after the cell has replicated its organelles, it moves into the S stage, ("S" standing for "synthesis.") This is the period in which the DNA is replicated using their origins of replication that were discussed in unit 1. Like the G_1 stage, the length of time that a cell spends in the S stage can be variable, although it is not nearly as extreme. By the end of the S phase, the genome of the cell has been copied. Two genomes are now present.

The cell is now ready to move forward into the next stage, G_2, also standing for "gap." In this stage, the cell makes last minute preparations for the much-anticipated eviction of the other genome, bringing us to mitosis.

Mitosis/Division

By the time the cell enters mitosis, the DNA has been replicated; each chromosome has been copied. Here is an image of a eukaryotic chromosome being replicated.

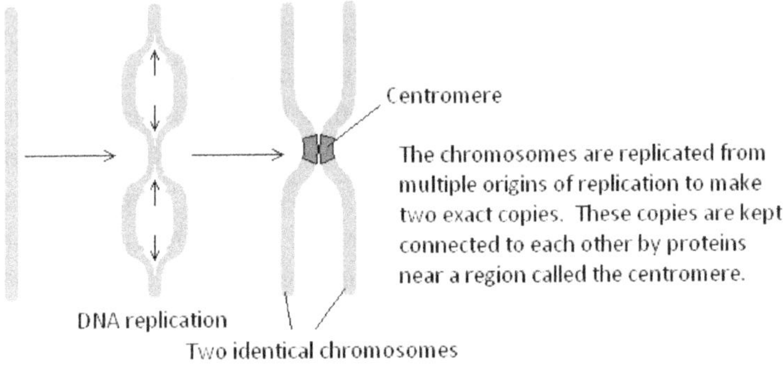

After the DNA is copied, each chromosome has a clone attached to it via a central region called the **centromere**.[16] The

[16] Though this forms an "X" like pair of chromosomes, do not confuse this with the "X chromosome." It is not the same thing. We will encounter the X chromosome in later sections.

centromere is a complex structure of proteins that forms around the same time as DNA replication. Its purpose is to keep track of the two copies; soon they will be going their separate ways, so the cell needs to keep close record of them. Let's see how the separation occurs. . .

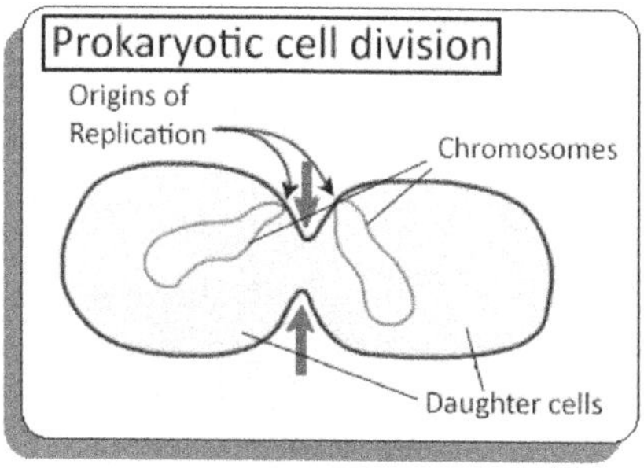

For prokaryotes, this is a simple process: the bacteria have their origin of replication machinery attached to the cell membrane near where the cell will split in half. The two new chromosomes will go their separate ways as the two new cells partition off their membranes (see picture to the left).[17] No centromere is necessary. Interestingly, this is also how mitochondria and chloroplasts segregate their chromosomes, lending more evidence to the theory that they were once free-living prokaryotes.

In eukaryotes, the process of DNA segregation is a little more complex and is broken down into several stages: **prophase**, **metaphase**, **anaphase** and **telophase**. Some books break these stages down into intermediate stages as well, but the general pattern still holds.

During prophase, two **centrosomes** appear near the nucleus and migrate to opposite sides of the cell. The centrosomes are connected to a large network of microtubules that span across the cell, and the star-like structures that they form are called **asters**. This network begins to align itself based on where the cell will soon divide. At this point, the nucleus slowly begins to disappear, and the

---

[17] I should point out that cell division in prokaryotes is not called mitosis. It is just called cell division. Mitosis is a eukaryotic phenomenon.

replicated DNA has been packed up very tightly, tight enough that it appears as discrete chromosomes under a microscope. There are twice as many chromosomes as usual, each one copied and attached to its clone via the centromere to form a number of little X figures. The following image is a simplification.

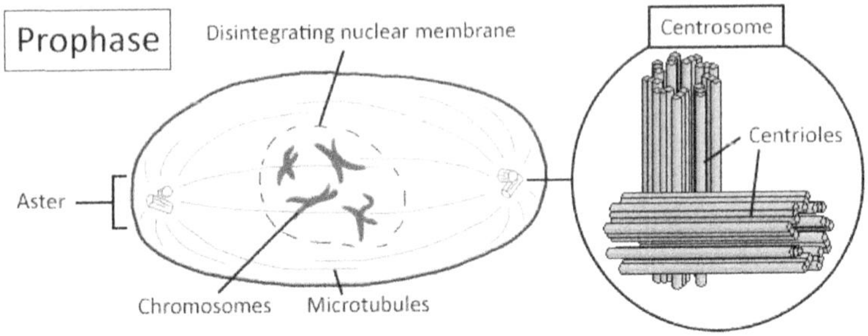

Near the end of prophase, little protein complexes called **kinetochores** form around the centromeres, (see image accompanying "anaphase" below). Once the kinetochores have attached to the centromeres, microtubules from the aster networks begin to connect to the kinetochores. After this connection has been established, the microtubule machinery is now ready to begin the next phase.

Next is **metaphase**. During this part of mitosis, the microtubule system pulls the chromosomes onto the line that will become the future boundary between the two new cells.

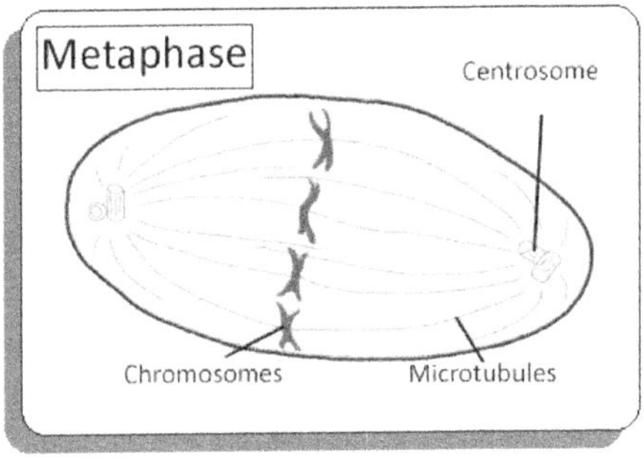

Next, **anaphase** starts when the chromosomes begin to separate from each other. During anaphase, the microtubule aster systems pull their respective chromosomes in opposite directions. The X-shaped chromosomes pull apart; the individual chromosomes go their separate ways. The next image shows kinetochores grasping onto the chromosomes as they are being pulled. It also shows a simple visual pneumonic for remembering anaphase.

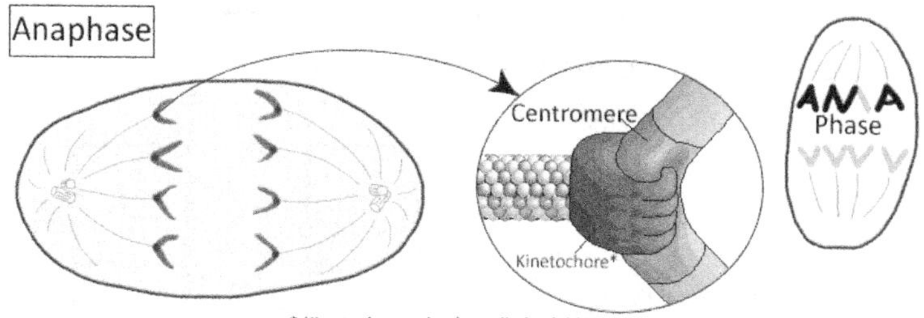

* Kinetochores don't really look like hands; I've drawn them like that to facilitate memory. Their actual structure is quite complex.

Last in the process of mitosis is **telophase**. In this stage, the microtubule aster systems begin to degrade and the cell (usually) starts to divide via **cytokinesis**. The cell nuclear membrane also begins to reemerge.

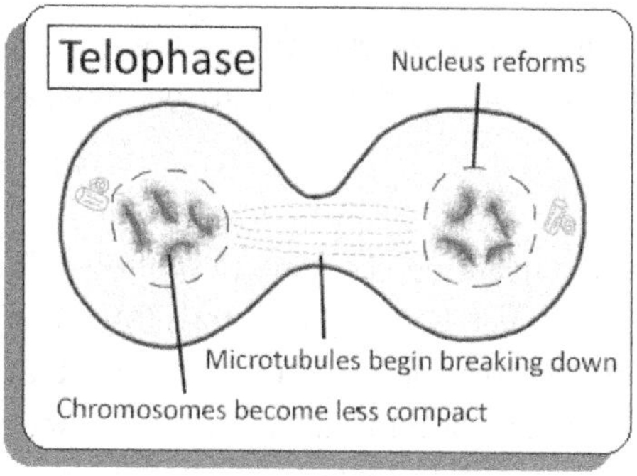

As it turns out, the process of cytokinesis is quite variable, depending on the cell type and species (see image on next page). In most animal cells, the cell membrane begins to pinch inward, slowly cutting the cytoplasm in half, roughly 50% for each daughter cell. In plants, the cells are surrounded by tough walls, and the division takes place by the gradual creation of a wall between the two cells. This wall begins as small islands along the new boundary, which slowly connect until the two daughter cells are finally cut off. In other species, such as algae, the cell wall is slowly built inward, analogous to the way the cell membrane splits off daughter cells in animals. In some protists and fungae the cytoplasm doesn't divide at all, in which case mitosis is referred to as **endomitosis** (mitosis without cytoplasm division). This results in a cell with multiple nuclei, called a **polynucleated cell** or **coenocytic cell**. Examples of this can also be found in some human cells, such as the megakaryocytes, (which make blood platelets), and skeletal muscle cells.

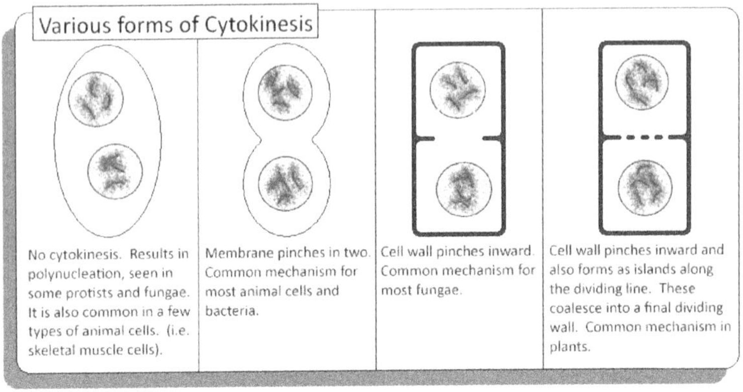

Regardless of the presence of cytokinesis, at the end of telophase the cell reenters interphase and returns to business as usual.

The Cellular Clock

At this point, a question might be occurring to you: How do the cells control the transition between G_1, S, G_2, or mitosis (M)? This question is an active area of research, primarily because cells that forget how to time these events can begin dividing rapidly, without restrictions. These we call cancer cells. In normal cells, the cell cycle appears to be governed by a protein called **cell-division-cycle** protein, or cdc. This particular protein guards a switch that moves the cell from one stage to the next. Cdc, however, has a baseline state of being inactive. Cdc doesn't lift a finger to do anything unless it has to, so the cell has a mechanism in place that periodically forces cdc to be active. As a cell progresses through the two G stages, it produces little motivator proteins called **cyclins**. There are two types of cyclins: **S-**

cyclin and **M-cyclin**. Throughout the G_1 stage, the concentration of S-cyclin slowly builds up until they force cdc from its lazy form to active form. This throws the switch in favor of DNA replication and causes a chain reaction that bumps the cell from the G_1 stage to the S stage. The S-cyclin is subsequently broken down, allowing cdc to be lazy again. At the beginning of G_2 stage, M-cyclin begins to build up. When M-cyclin reaches a critical concentration, they cause cdc to become active again, pushing the cell into mitosis. The M-cyclin are then broken down during the course of mitosis and cdc once again becomes lazy. That in a nutshell is how the cell cycles through its divisions, by slowly making S-cyclins and then M-cyclins that keep cdc on its toes.

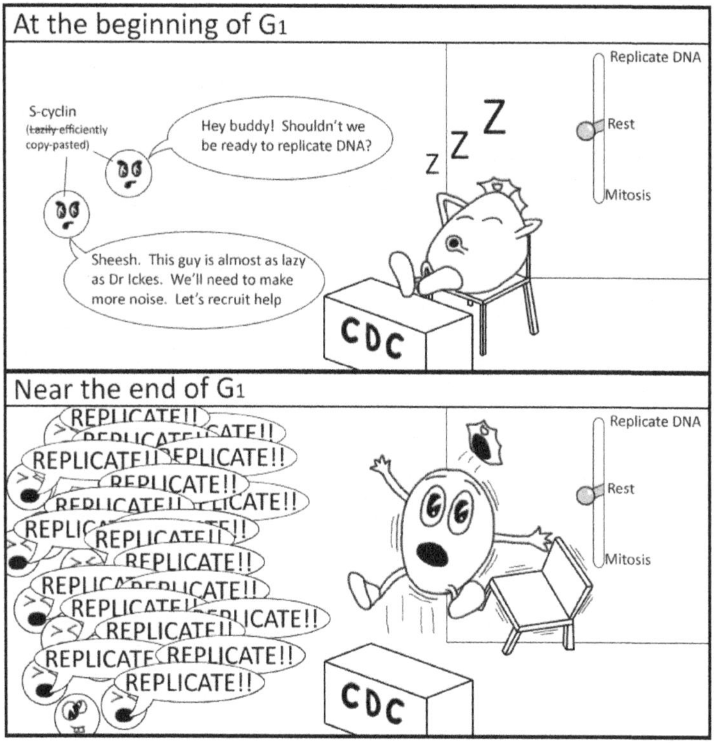

In this cartoon, we see two S-cyclin proteins grumbling because they cannot get cdc to throw the cell into DNA replication (S stage). Near the end of G_1, the number of S-cyclins has become so high that the cdc is alerted to throw the switch. A similar cartoon could be drawn for the transition into mitosis. M-cyclins would slowly build up until they become numerous enough to wake cdc, which would throw the switch to enter mitosis.

Summary

In this section, we discussed the cell cycle, including mitosis and interphase.

- Normal eukaryotic cells alternate between interphase and mitosis. The amount of time that is spent in the individual stages varies. Some eukaryotic cells, such as muscle and nerve cells, stay in interphase indefinitely.
- Prokaryotic cells also alternate between interphase and a mitosis-like stage, called (quite simply) cell division. Their division is typically much simpler than eukaryotic mitosis.
- Interphase is divided into G_1, S, and G_2 stages.
- During G_1, or "gap 1" stage, the cell replicates various organelles and conducts normal business of protein production.
- During S stage, or "synthesis" stage, the cell replicates its DNA in preparation for mitosis.
- During G_2 stage, or "gap 2" stage, the cell completes final errands necessary to enter mitosis.
- Mitosis is divided into four substages: prophase, metaphase, anaphase and telophase.
- During prophase:
    - 1) Two centrosomes (each containing two centrioles) migrate to opposite sides of the cell and serve as centers for the formation of a cell-wide microtubule system. These star-like microtubule centers are called asters.
    - 2) The nuclear membrane disappears.
    - 3) The DNA begins to condense into discrete chromosomes. Each chromosome is attached to its duplicate via proteins connected to their centromere region.

    - 4) Kinetochores form near each centromere, and the microtubules radiating out from the asters attach to the kinetochores.
- During metaphase, the chromosomes are lined up along the middle line.
- During anaphase, the chromosomes detach from their duplicates and are pulled by the microtubule/kinetochore structures toward opposite asters.
- During telophase:
    - The chromosomes begin to unpack, making them seem less discreet.
    - The microtubule aster system breaks down.
    - The nuclear membranes of the two new nuclei begin to form.
    - Cytokinesis (if present) divides the cell.
- The cell alternates from G_1 to S and from G_2 to M by making little proteins called cyclins. These cyclins build in concentration until they cause a protein complex called cdc to start a chain reaction that changes the cell's stage in the cycle. S-cyclins cause cdc to switch the cell into S stage. M-cyclins cause cdc to switch the cell into M stage. Once these stages are accomplished, the respective cyclins are destroyed, and the cell reverts to the next G stage.

Unit 7

The Cellular Basics of Sexual Reproduction

In this section we cover:

1) Homologous Chromosomes
2) Meiosis
3) The Diversity resulting from Sexual Reproduction
4) Life Cycles

In the last unit, we had a look at a cell's typical day. It spends part of the time doing business as usual and part of the time dividing in a process called mitosis. For any purely **asexually** reproducing eukaryotic species, mitosis is essentially the end of the story for cell division. For **sexually** reproducing species, such as humans, there is a little more to the saga. As it turns out, there is a small population of cells within your body whose business it is to divide in a very *different* way. These particular cells are involved in the production of **gametes**, the sex cells, and the way they divide is called **meiosis** (pronounced "my oh sis."). This particular process shares many features with mitosis but has a few important differences as well. A study of these differences will lead us to a vastly higher understanding of biology in general, because the process of meiosis is common to a great many organisms on this planet, (not just humans), and is an important part of the stage upon which we study heredity and evolution. To tackle the learning of these topics, we'll need to back up a little with a quick discussion of the genome and how it's organized.

The Homologous Chromosome

For all practical purposes, eukaryotic cells can be divided into two groups, **haploids** and **diploids**[18]. Diploids have a special feature built into their genome: **homologous chromosomes**. These chromosomes are paired, each having similar gene layout and function. Let's learn by example. Suppose we are examining a diploid genome, and one particular chromosome in it has a gene starting at nucleotide 10,355 and is 1200 nucleotides long. Further suppose that the protein made from this gene affects the color of the organism's eyes, say making them blue. The homologous partner chromosome would *also* have a gene positioned at nucleotide 10,355 that is 1200 nucleotides long, and the gene at this location would make a protein that also affects the color of the eyes, although the color does not have to be the same. For example, the eye color from the

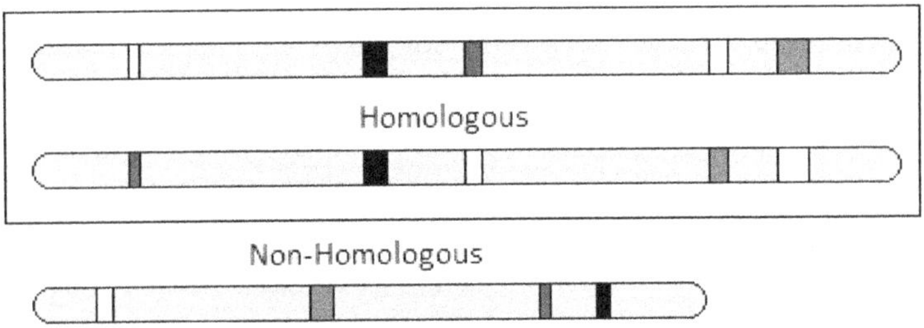

homologous chromosome might be brown. Naturally, you may wonder which chromosome gets to color the eyes, which is an excellent question. We'll be covering that topic in the upcoming chapters on heredity in the next topic text. For now, the important thing to understand is that in diploids 1) there are two genes for every

---

[18] Technically there are also triploids, quadruploids, and others, but these are *way* outside the scope of this book and are infrequent enough to be viewed as the exceptions to a general rule. Regarding these exceptions now would only serve to guarantee confusion.

function, and that 2) at every point where there is a gene on one chromosome there will be a gene at the equivalent point on the partner chromosome. Haploid cells do not have the “extra” genes.

Here are sample genomes from generic diploid and haploid organisms.

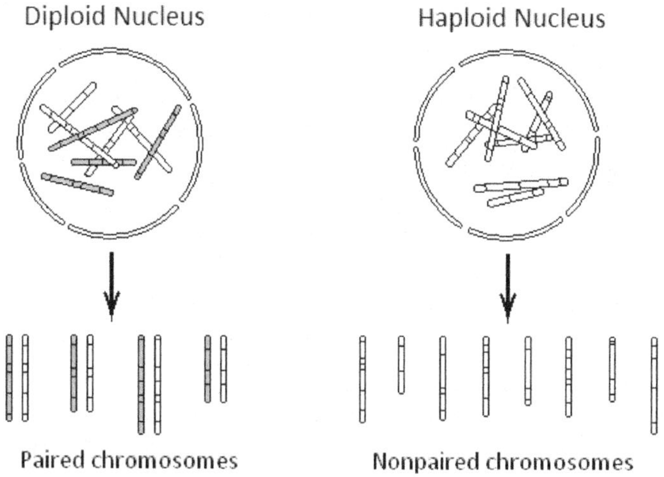

(The coloring is purely for illustrative purposes. Chromosomes do not have color.)

So far the punch-line is simple: in diploid organisms each chromosome has a partner that is strikingly similar. This little feature lies at the heart of the differences between meiosis and mitosis. If you’ll recall, during mitosis the replicated chromosomes remain attached to each other by their centromere regions, forming little X-structures. The purpose is to ensure that both chromosomes go their separate ways into the two new nuclei. In meiosis, as we will see, four new cells are created through two division cycles, and a similar strategy of keeping chromosomes attached to each other is employed to ensure that each of the replicated homologous chromosomes go their separate ways.

To facilitate this goal, the replicated chromosomes in the X-structures[19] are saddled up next to each other to form a **synaptonemal complex**[20], as shown below. Meiosis proceeds with two mitosis-like rounds of division. During the first division, the synaptonemal complex is split down the middle, so that each X-structure goes its separate way. During the second round of division, the individual X-structures are split, just as they are in mitosis, so that the individual chromosomes go independently. When all is said and done, the little trick assures that each homologous chromosome has now been sent to a different new cell.

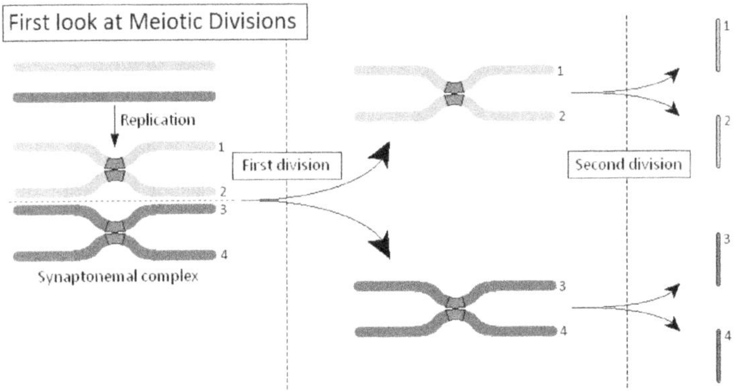

As it turns out, there is more to the story . . . a very important part, in fact. During the construction of the synaptonemal complex, the homologous chromosomes will often switch parts of their DNA with each other. In this process, the DNA of the two chromosomes line up perfectly, then switch parts as illustrated by the following graphic.

---

[19] Once again, please do not confuse this with the "X" chromosome. The X chromosome is unrelated to this particular phenomenon. We will learn about the X chromosome later.

[20] There is a great deal of protein involved in the formation of the synaptonemal complex. These are not shown in the images to maintain simplicity.

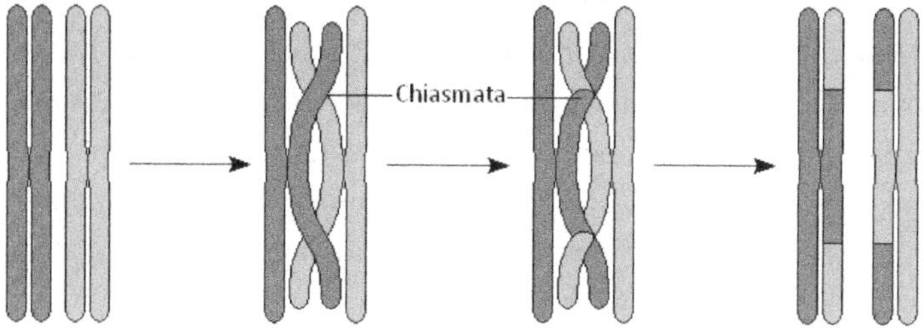

This is called **crossing over** or **recombination**, and the actual spots where the chromosomes cross are called **chiasmata**. Generally, chiasmata usually occur between gene locations and not within them. In other words, DNA typically does not cross over right in the middle of a gene, but rather in a junk region.

Taking recombination into consideration, an updated view of the meiotic division process would look like this:

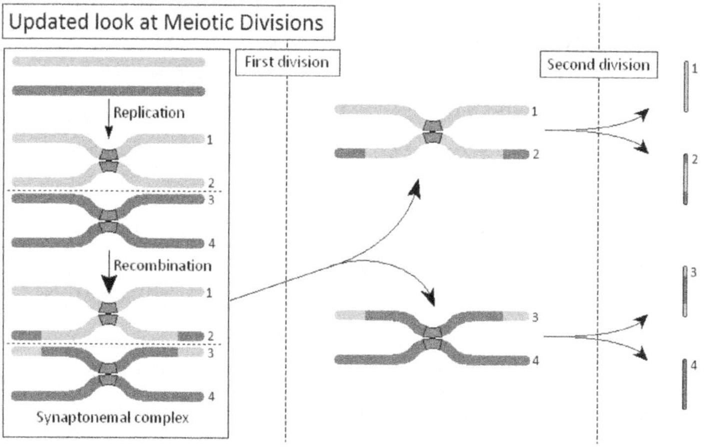

The above images show only the nearest two homologous chromosomes exchanging the tips of their chromosomes. The light gray and dark gray chromosomes on the outside of the synaptonemal complex can recombine with each other as well. Taking all this into account, the number and positions of the recombination events is variable and essentially impossible to predict. As such, the above image represents only one possibility among the nearly limitless number. Here are a few more possible outcomes:

Examples of other possible recombinations. . .

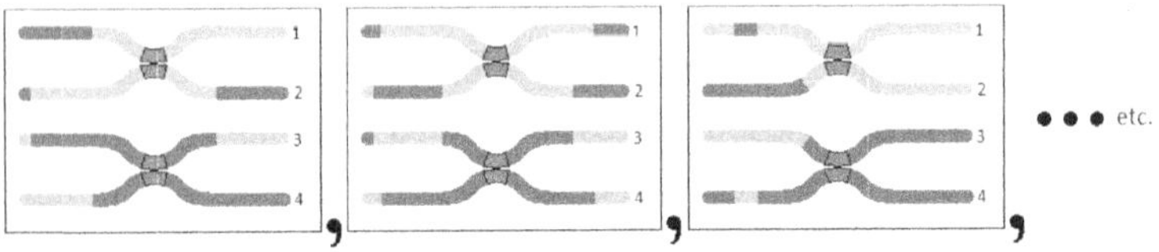

As you can see, there are many ways to duplicate, recombine and separate homologous chromosomes, and as we'll discuss more near the end of this unit, it leads to an unbelievable spectrum of diversity amongst a population of individuals. Before we pick that discussion back up, however, let's have a closer look at meiosis itself.

Meiosis

In the last chapter, when we studied mitosis, we saw that it involves four broadly defined stages: prophase, metaphase, anaphase and telophase. Meiosis follows a similar pattern, but division happens *twice*. It starts with one "parent" cell that has duplicated its chromosomes, and undergoes *two* mitosis-like divisions to create four "child" cells. The business of homologous chromosome recombination takes place during the first division. The combined process is shown on the following page.

As that image shows, the process starts with a diploid cell that has replicated its chromosomes. These chromosomes undergo recombination around the time of metaphase I. After the completion of telophase I, the cell has split into two smaller cells. Though some duplication and recombination has taken place, both cells basically have only one representative from each of the original homologous pairs. Each exists in their duplicated state and tied to their duplicate via their centromere. After a brief intermission-like **interkinesis** stage, (during which the centrioles and other organelles replicate), the second division proceeds and these duplicated chromosomes go their separate way. By the end, the process results in four child cells, which then go on to further modification to become gametes.

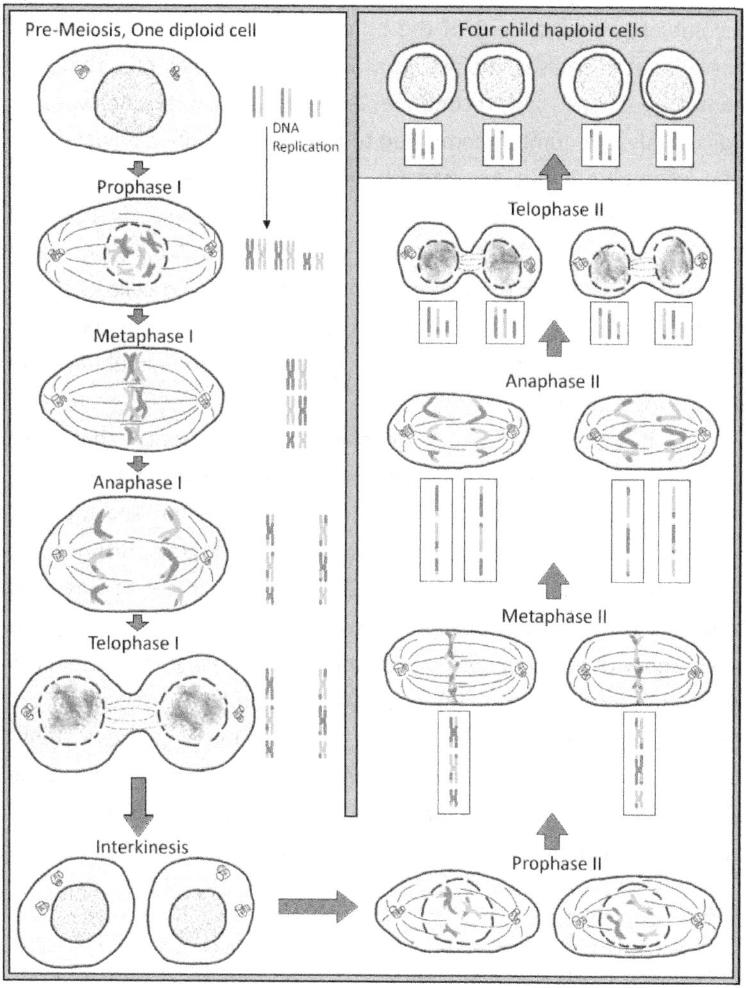

Diversity of Gametes

As we saw earlier, each individual homologous pair can recombine a vast number of ways, creating a nearly endless spectrum of possible outcomes. We considered only a few of these possibilities before we looked at the picture of meiosis. Diploid cells typically have many more than just one homologous pair, however, and each pair is recombining independent of the others. That means the number of possibilities increases many-fold. Suppose, for example, that one homologous pair can recombine 1,000,000,000 different ways (a ridiculously low number compared to the *actual* number of ways they can recombine – but no matter). If a second pair can recombine 2,000,000,000 ways, the two pairs can recombine a total of 1,000,000,000 x 2,000,000,000 = 2,000,000,000,000,000,000 ways. That is only with two homologous pairs! Many diploid cells have ten or more homologous pairs. Humans have 23 homologous pairs! At this point, we have a massive number of ways we can recombine DNA. It gets worse though. . .

If you look back at anaphase I, you'll notice that each homologous pair splits apart independent of each other. For example, the dark gray chromosome from the first pairs goes left and the light gray one goes right. In the second pair, the dark gray chromosome goes right and the light gray one goes left. The number of ways this **independent assortment**[21] of chromosomes can occur is 2^X, where X is the number of homologous pairs the diploid cell has. If the diploid cell were from a human, the number of ways would be 2^{23}. The statistics technically get more complex as we go, since many of these "new" combinations are already covered with our first set of numbers, so we'll leave it to the mathematicians. When all is said and done, however, the point is that you can make an essentially limitless

---

[21] The fact that each chromosome pair separates independent of each other is called the Principle of Independent Assortment

number of possible gametes. The odds are so staggering that in all likelihood NO two gametes will be the same.

When reproduction occurs, one gamete from a male organism (usually called a **sperm**) will join with one gamete from a female organism (usually called an **egg**). If both the male *and* female can each produce a practically limitless combination of gametes, the total possible number of ways to combine the two gametes is squared. At this point, the argument has hopefully been driven straight home: No two siblings are identical.[22] This is the advantage of sexual reproduction. The number of possible outcomes is vast. If organisms only reproduced by mitosis, they would only replicate the same old boring genome each time. What happens if the environment changes and that particular genome is no longer able to keep up? The entire line or organisms might die off.

That is not to say that some organisms don't survive by asexual reproduction. Bacteria are notorious for reproducing asexually. They just divide in half and do it so rapidly that they flood the environment with themselves. Take your hand and touch practically any object near you; the odds are good you just touched a LOT of bacteria. Bacteria populate areas very easily, but die just as easily too. Take an antibiotic and bomb them with it, and more often than not the entire colony will get wiped out. Rapid clone-producing division grants the advantage of rapid environment invasion, but the

---

[22] What about twins? Well, twins, triplets, quadruplets, etc., result from abnormal divisions of the fertilized egg. This takes place *after* the sperm has already united with the egg, which means that each duplicate individual receives the same combination of DNA. This is the reason all such siblings are "identical." Otherwise, for standard siblings, the odds of two being the same are so remote that it is pointless to even consider. On the other hand, I feel inclined to point out that twins, triplets, quadruplets, etc. are not technically identical either – only their DNA is. If you take microscopic samples of any tissue from their bodies and compare them, you'll find vast differences. On the macroscopic level, they look practically identical, but that is because our vision is not very discriminating.

disadvantage of being poorly adapting when the environment changes.

Let's take a moment to toss the effects of mutation into this argument. As we saw above, meiosis usually does not swap DNA in the middle of genes. As a result, meiosis shuffles around existing genes, but usually does not create new ones. Mutations introduce the changes required to make new genes. Now, if we add in the basic rate of mutation that we saw several units ago, (the one in a million nucleotides per division), we increase the genetic variability even more. Therefore, sexual creatures achieve incredible variability from meiosis and mutations, whereas organisms such as bacteria rely mostly on mutations for their variability, (although in the next chapter we'll look at a few other mechanisms they use to get around this limitation). For now, let's look at an example of meiosis.

Example of Meiosis: Mammals

It may be instructive to consider the most obvious example of meiosis, that which occurs in human beings. The process in humans is sufficiently similar to many animals (especially mammals) that it can be widely generalized. In males, meiosis occurs in the **seminiferous tubules** of the **testis** (i.e., **testicles**), and the result of this meiotic process is the **sperm** cell. The generation of sperm cells is called **spermatogenesis** and begins with a diploid **primary spermatocyte**. The first round of division, (meiosis I), produces **secondary spermatocytes**. Meiosis II produces four haploid **spermatids**, each with a completely unique DNA combination. These spermatid cells mature into **spermatozoa**. They have a long flagella tail, a single mitochondria wrapped around the junction of the tail, and a cell body with a special cap called the **acrosome**. This cap is full of enzymes that help the sperm to penetrate the egg. The process of spermatogenesis occurs from the onset of puberty and continues until death.

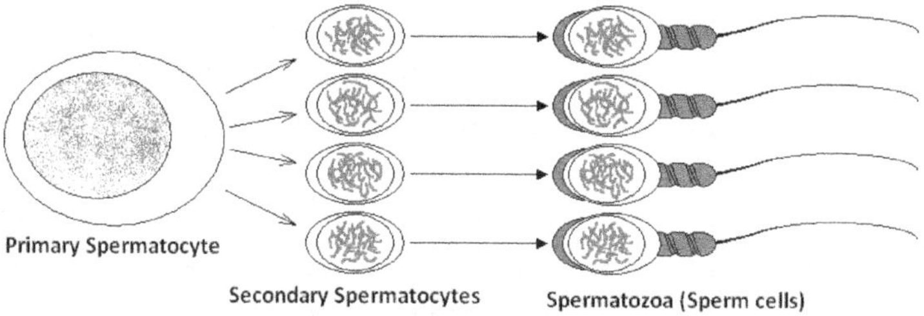

In females, meiosis occurs in the ovaries and begins shortly after their formation in utero (i.e., while still a fetus in the womb). The initial cells that start the process are called the **primary oocytes**, which are diploid. They pause just after completion of the prophase I, and remain at that step until the onset of puberty. At that time, each month one primary oocyte within the two ovaries is chosen alone to complete meiosis. The chosen cell then resumes meiosis I, but unlike spermatogenesis, the resulting daughter cells are not the same. One cell is called the **first polar body**, while the other is the **secondary oocyte**. The secondary oocyte continues through meiosis II, once again producing two uneven daughter cells. One is called the **ootid**, which ultimately matures into the **ovum** (egg). The other cell becomes a **second polar body**. The first polar body sometimes completes meiosis II as well, becoming two secondary polar bodies.

This entire process of ovum production is called **oogenesis**, and is outlined here:

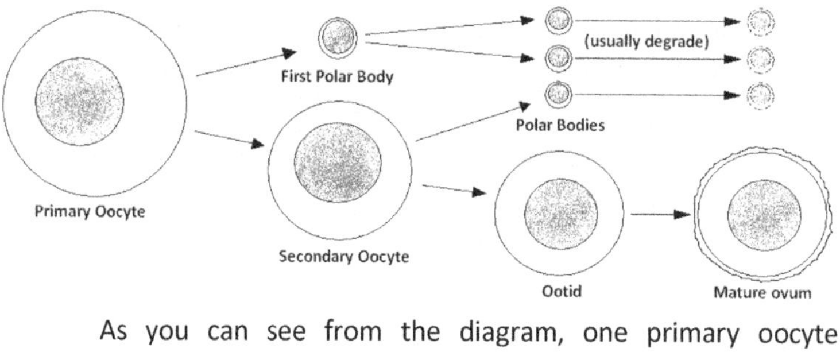

As you can see from the diagram, one primary oocyte produces one ovum and three secondary polar bodies. Why doesn't it just make four oocytes? The answer to that is a practical one. Fertilization occurs when the sperm enters the egg, and the first few events that follow take place within the volume of the egg. Eggs, therefore, need to have as much volume and material as possible to maximize the chance that fertilization will succeed. So, one egg ends up hogging all the stuff from its runt sister cells, who each become tiny secondary polar bodies. There doesn't appear to be much purpose of polar bodies, and they usually disintegrate after a short time. That's life, I guess. The favorite sibling gets all the inheritance.

Fertilization and Life Cycles

When the sperm and egg unite, their haploid nuclei combine to create a new diploid nucleus, marking the start of an organism with genetic information that differs from that contained in the cells of either parent.

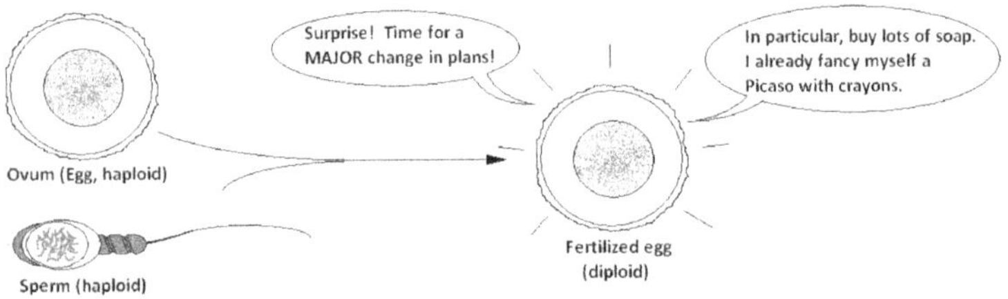

We are now prepared to examine a more global understanding of life by realizing that it is essentially the continuation of a cycle. Life within the human species – like most animals – is spent primarily in the diploid "individual" form. A small part of the time is spent in haploid form as gametes. In many advanced plants, the pattern is the same. Comparing animals and advanced plants with progressively more primitive plant forms, we could draw the life cycles like this:

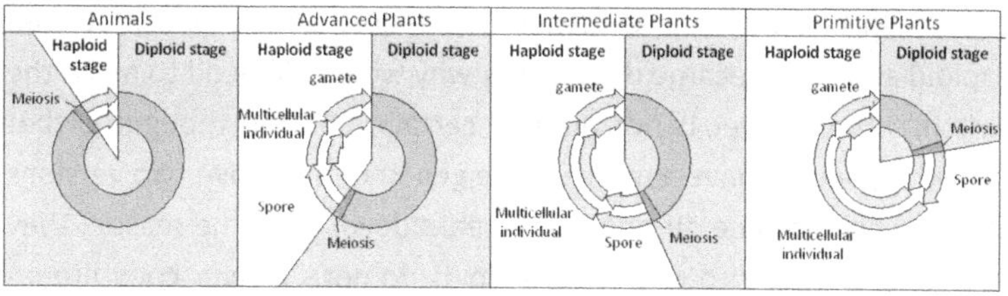

The trend for plants is rather clear: the further you go back, the more dependence there is on the simpler haploid form of existence. How do we know this, and does the trend exist for animals? Well, we know about intermediate and primitive plants because of similarity with fossils. For example, ferns are not much different than their fossil forms; they've been around with little variation for almost 400 million years. The modern day ferns live about half their lives as small multicellular haploids, called **gametophytes**. These create gametes, which then join to make the diploid fern. The fern then creates spores through meiosis, and the spores reform the gametophytes. At that point, the cycle begins again. The same thing is true for primitive plants. A few representatives are still around today, and they spend very little time as diploid individuals. So, what about animals? Well, practically all animals have diploid life cycles. They partake in a brief haploid stage for the purpose of reproduction, and that is all. Even the reptiles and fishes, arguably similar to their fossil representatives, live their lives mostly in diploid form. On the other hand, if you go back far enough – before the dinosaurs, before things walked on land, before fish in the sea, you see fossil forms that get progressively smaller and smaller. The trail of plants and animals begins to converge into simpler paths that are not unlike each other. If plants were mostly haploids back in those days, one can guess that animals began with similar life cycles as well. So, in a global sense, the pattern appears to hold. Life started simple and slowly got more complex.

In line with the evidence, the theory goes that life started haploid and later became diploid, but why would life need to make the switch? The answer is not entirely certain, but one thought is that diploid organisms have a more secure genome; they have two versions of each chromosome. If one copy breaks down for some reason, they have another copy to rely on. Haploids do not. If their copy breaks down, they are done.

Plants have an easier form of life than animals do. After all, they do not need to be concerned with all the thermodynamic and engineering difficulties associated with movement. They don't need heightened senses in order to detect resources and avoid getting chased down by some nasty critter. Plants just sit there, soaking minerals from the ground and catching rays from the Sun. If an animal eats some of their leaves, it's no big deal; they just make more leaves. In that sense, it isn't hard to see why some forms of plants are still around that can make it as haploids. With animals, however, the slightest genetic problem can create all kinds of havoc. It definitely helps to have an extra version of your chromosomes hanging around.

Summary

In this section, we discussed meiosis.

- A homologous chromosome is one that has a partner chromosome that has a similar genetic layout and length. The gene appearing in a particular position will have the same general purpose, (i.e. eye color, hair texture, whatever), in its partner chromosome. While the homologous genes have the same general purpose, they do not always result in the same effect. For example, one might influence the eyes to be blue. The other might influence them to be green. How color is decided is the topic of future chapters.
- Non-homologous chromosomes have completely different genetic plans. Their genes do entirely different things and are positioned at different places.
- Organisms whose chromosomes all have homologous partners are called diploids. By comparison, organisms with no homologous chromosomes are called haploids. Diploid genomes are believed to be more stable because they have two "versions" of each chromosome.
- Meiosis is a cell division process that results in four unique daughter cells. It takes place in two distinct mitosis-like divisions.
- During the first division of meiosis, specifically around metaphase I, the homologous chromosomes pair up in synaptonemal complexes and swap portions of their DNA. This is called crossing over or recombination. By the end of telophase I, the homologous chromosomes are each separated into two new daughter cells.
- During meiosis 2, the duplicated DNA chromosomes are segregated into four total haploid child cells. In the male, these four cells mature into sperm cells. In the female, only

one of these four matures into an egg. This egg inherits the majority of the original cytoplasm and materials. The other three sister cells, called polar bodies, inherit very little cytoplasm. The polar bodies eventually die off, with little known purpose.

- The number of ways that DNA can be recombined during meiosis is vast. This contributes to the wide variety of genetic combinations found in sexually reproducing species. Asexual organisms do not benefit from this genetic variation, and depend on mutations and a few other minor mechanisms to increase their genetic variation.
- After the sperm and egg unite, the resulting cell becomes a new diploid individual, and for the initial stages that follow, this new individual stays the same size as the egg, utilizing the materials inherited by the egg.
- Most modern plants and animals spend the majority of their life cycles as diploid organisms, with very little time spent in the haploid form. The more primitive forms of plants spend a larger amount of time as haploids, so that they can appear as two entirely different physical forms. Ferns are an example, in which their haploid stage is referred to as a gametophyte. Primitive animals are also believed to have spent most of their life cycle in the haploid form.
- Diploid organisms are believed to have a more robust genome because each chromosome has a homologous version. If one gene experiences a mutation (i.e. breaks), the other version of the gene located on the homologous chromosome is still there to carry on the function. The theory is that life gradually shifted from a haploid form to diploid to take advantage of this extra protection.

Unit 8

Additional Mechanisms for Genetic Variability

In this section we cover:

1) Bacterial transformation

2) Bacterial Conjugation

3) Bacterial Transduction

In the last unit, we touched on the amazing variability that arises from sexual reproduction. In this unit, we'll begin to look at how variability is accomplished in bacteria. Specifically, we will turn our attention to three bacterial processes: **transformation**, **conjugation**, and **transduction**.

Transformation

Many species of bacteria are capable of taking up fragments of DNA that are floating around. After acquiring the new DNA pieces, the bacteria incorporate them into their own genome, and if the pieces contain genes, the bacteria gain the benefits of the new genes. This is called transformation, and the most obvious example, (and possibly the most relevant to us as humans), is when they pick up genes that give them immunity to an antibiotic. This process is especially obvious when you add the dead remains of antibiotic-resistant bacteria to a culture of living bacteria that are *not* resistant. Some portion of the non-resistant bacteria will pick up the DNA pieces from the dead bacteria and become resistant to the antibiotic! To illustrate how remarkable this is: it would be like a person catching the DNA from a feather and gaining the ability to fly. Transformation, this is one of many reasons why antibiotic resistance arises.

Note that not all bacteria can pick up and incorporate external DNA, but many can be induced to be able to do it. Recall that many bacteria have cell walls that protect them from the environment. In order for these armored bacteria to pick up stray DNA, you need to treat them with chemicals that break down the cell wall first. This, however, is a bit of a trick, and these kinds of bacteria are much less likely to be transformed in natural settings without our laboratory tinkering.

Bacterial Transformation

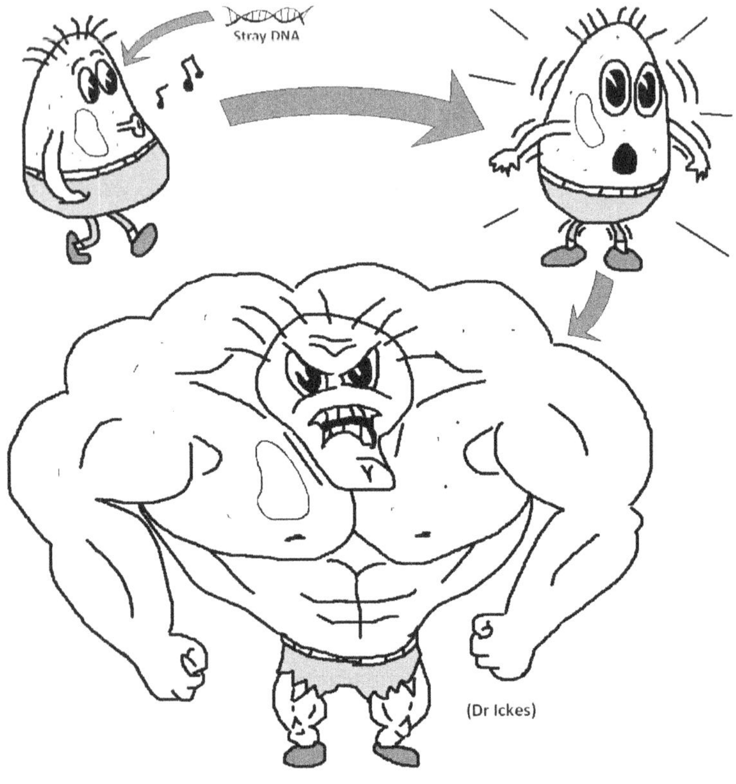

Conjugation

Bacterial cells do not have sex. Well, at least they don't have sex in the form that we recognize. As it turns out, some bacteria undergo a process called conjugation, the closest thing to sex happening in their little world. During this process, one bacterium pulls up next to another and extends a bridge of cytoplasm (called a **pilus**), like a pirate ship broadsiding another ship, or one cell sticking a straw in another. Once the pilus is established, bits of DNA are transferred from the pilus-forming bacteria. The odd thing is, it's usually not the bacterial chromosome that gets transferred but rather little loops of DNA called **plasmids**.

Bacterial Conjugation

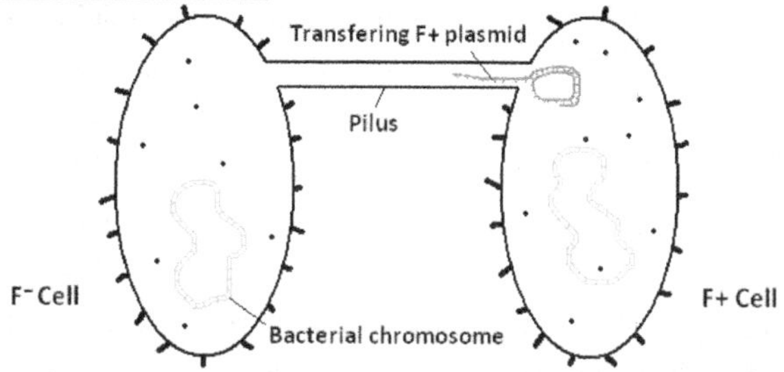

If you'll recall from our discussion of cell anatomy in Topic Text B, plasmids are small rings of DNA that are very common amongst bacteria. Quite often, they carry genes that give the bacteria chemical upgrades. For example, one plasmid might confer the ability to breakdown a normally-harmful chemical in the environment, giving the host bacteria an extra boost in an otherwise poisonous world. Another plasmid might contain the genes for synthesizing chemicals that are vital, whereas the bacteria would otherwise just have to rely on these chemicals being premade in the environment.

In short, plasmids are tiny upgrades, and conjugation is one way the bacterial world passes them around. There is an important hurdle to this free-trade of genes, though: only bacteria having the genes for making the **pilus** can initiate conjugation. This particular ability comes encoded in what is called the **sex factor** (or **F factor**, for "fertility"), which is the set of genes that work to make the pilus and all other proteins necessary for the genetic transfer. Bacteria that have these genes are called F+ cells and are considered "male." Those without are designated F- cells and considered to be female.

After pilus formation, the DNA that contains the sex factor genes are copied and passed to the F- cell. At that point, the F- cell also becomes F+ and can initiate conjugation with other F- cells. Naturally, one might think that before long all bacteria would become F+, but that isn't the case. As it turns out, it takes long enough to efficiently make a pilus that F- cells have plenty of time to divide, so that in many cases the proportion of F+ to F- cells in a population can actually drop. So many F- cells, so little time. . .

There is a theory that proposes that the sex factor genes are a kind of virus, and that bacteria do not conjugate deliberately. Rather, the sex factor uses the bacteria to form the pilus, which acts as a stationary "viral" capsid of sorts. Once the sex factor has invaded a new cell, it prepares itself for another cycle, invading more and more bacteria over time. In other words, the sex factor genes exist purely because they can and for little other reason than that.

The important consequence to this process is not just the replication and propagation of genes that think only with their pilus, but that *other* genes can sometimes be transferred along with the sex factor genes. Through conjugation, bacteria can spread new DNA that confers unique immunities to antibiotics or new abilities to synthesize chemicals. So no matter how inconvenienced the bacteria might feel about the whole thing, conjugation carries the added benefit of receiving new genetic information.

Transduction

There is one more major way in which bacterial cells acquire new genetic information: through viral infections, a process called transduction. Recall from several units earlier that when a virus invades a cell, it uses the cell's protein-making machinery to manufacture new capsids. These capsids act as escape pods for the virus, taking up the viral genetic information and heading off to a new cell. Sometimes, however, other DNA, (like the little circular plasmids, for example), are taken up by the virus by mistake. When this happens, the viral capsid takes off for a new cell and inserts this "bystander" DNA into the new cell. In this manner, viruses help to spread genetic information throughout a bacterial colony.

When it comes to bacteria, there is one particular virus that scientists use a lot to deliberately insert genetic information into bacteria cells. This particular virus is called a bacteriophage, and it looks like a little spider. Basically, scientists have developed techniques to insert various plasmids into the capsids of these viruses, then add the plasmid-containing capsids to a population of bacteria. The capsids then inject the new genetic material into the bacteria. Depending on the plasmid used, the recipient bacteria may gain a new ability or resistance, and scientists can use that to study the bacteria or the genes they inserted.

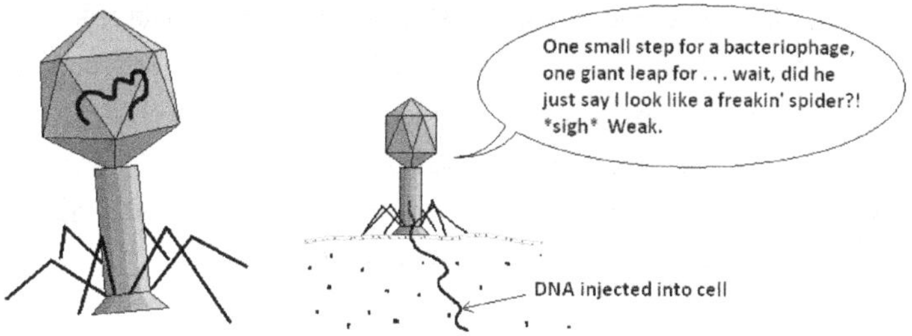

Summary

In this chapter, we discussed three major ways in which bacteria acquire new DNA.

- The process of bacterial transformation occurs when bacteria pick up stray DNA that is floating around. If this DNA contains genes for useful proteins, the bacteria gains whatever abilities the proteins grant. Transformation can make bacteria newly immune to an antibiotic, or give the bacteria the ability to synthesize chemicals that it could not previously make.
- Bacterial conjugation is the closest thing to sex that bacteria have. It begins with an F+ cell (a cell having the sex factor genes). This cell creates a straw-like pilus that bridges the gap between it and another cell. The other cell is called the F- cell. Once the pilus attaches, DNA (usually containing the sex factor genes, as well as other stuff) is copied and transferred to the F- cell. After the F- cell acquires the sex factor genes, it becomes an F+ cell and can initiate conjugation with other bacteria cells. Like transformation, this process can make bacteria newly immune to an antibiotic or be able to make new chemicals.
- The DNA that is transferred during conjugation is usually a plasmid.
- Transduction is the process in which a viral capsid accidentally injects non-viral DNA into target bacteria. This DNA can consist of gene fragments or an entire plasmid. Either way, like transformation and conjugation, it can result in bacteria gaining new abilities.
- A common bacterial virus often used in labs is the bacteriophage, which looks like a little spider.

Index

P

R

S

T

U

V

W

X

Made in the USA
Monee, IL
04 November 2026

67377799R00059